Systems Engineering

Systems Engineering

Fifty Lessons Learned

Howard Eisner

CRC Press
Taylor & Francis Group
Boca Raton London New York

CRC Press is an imprint of the
Taylor & Francis Group, an **informa** business

First edition published 2020
by CRC Press
6000 Broken Sound Parkway NW, Suite 300, Boca Raton, FL 33487-2742

and by CRC Press
2 Park Square, Milton Park, Abingdon, Oxon, OX14 4RN

Library of Congress Cataloging-in-Publication Data

Names: Eisner, Howard, 1935- author.
Title: Systems engineering : fifty lessons learned / Howard Eisner.
Description: First edition. | Boca Raton, FL : CRC Press, 2020. | Includes bibliographical references and index.
Identifiers: LCCN 2020013254 (print) | LCCN 2020013255 (ebook) | ISBN 9780367422424 (hardback) | ISBN 9781003002505 (ebook)
Subjects: LCSH: Systems engineering--Management.
Classification: LCC TA168 .E3874 2020 (print) | LCC TA168 (ebook) | DDC 620.0068--dc23
LC record available at https://lccn.loc.gov/2020013254
LC ebook record available at https://lccn.loc.gov/2020013255

ISBN: 978-0-367-42242-4 (hbk)
ISBN: 978-1-003-00250-5 (ebk)

Typeset in Times
by Deanta Global Publishing Services, Chennai, India

This book is dedicated to my wife, June Linowitz, a professional artist who is devoted to creating art for the world and the enjoyment of her friends, colleagues, and family. It is also dedicated to my children and their spouses, who are Susan Rachel Eisner Lee, Oren David Eisner, Joseph Lee, and Tara Burke. Further dedications are to Oren's sons Zachary and Ben, deceased (Seth) son Jacob, and Susan and Joseph's children Gabriel and Lee.

Contents

Preface

On page 9 (Table 4.1) of the fourth edition of the *Systems Engineering Handbook* (produced by INCOSE), it is noticed that an important date in the origins of systems engineering (SE) as a discipline was cited as 1937. That was two years after my birth year, which makes me older than the entire field of SE. The handbook also takes note of other significant dates such as:

1954 – the RAND Corporation recommending and using the terms (systems engineering)
1962 – Hall's classic book on systems engineering
1990 – the formation of NCOSE (National Council on Systems Engineering)
2008 – ISO, IEC, IEEE, INCOSE, and PSM harmonize SE concepts in standard 15288:2008

This all prompted me to pause for a moment or two and think about systems engineering, a field in which I have worked for some 50 years, starting from the age of 25–75 (I am now past 80). This involvement intersects with my 30 years in the industry as a working engineer, manager, executive, and president of two high-tech companies. It also intersects with my 24 years as a professor at the George Washington University, where I was a strong advocate for the study and use of systems engineering. "Time to sum up", I thought. And that is what this book is all about. I sat down and decided to document the "50 lessons learned" over this time period. What were some messages I could leave for the next couple of generations to contemplate with the field they chose to work in?

So here they are – 50 of them – in six categories and chapters:

- Technical: Chapter 1
- Management: Chapter 2
- Idea-Based: Chapter 3
- People-Oriented: Chapter 4
- Miscellany: Chapter 5
- Top Ten Lessons: Chapter 6

I hope that readers will indeed find them interesting and useful in their lives as systems engineers. Some lessons are more powerful and long lasting than others, and these are in my "top ten" list, which is Chapter 6. Readers may wish to pause to think about this set of ten lessons to see if there is any resonance with their experiences. In any case, it's all about thinking and re-thinking, reviewing and re-evaluating, which is what systems engineers like to do.

This book is a retrospective from a systems engineering perspective. As such, it looks back at 50 years of working in the field of systems engineering and cites some 50 "lessons learned" during this rather long period of time. These lessons are organized into six categories and chapters, as mentioned above.

This author, from approximately age 25 to 75, has worked on many problem areas associated with quite a few clients. These have included investigations in the domains of:

- Satellites and related ground systems
- Aircraft and aviation
- Air traffic control systems
- Cost-effectiveness evaluations
- Systems architecting
- Information systems
- Torpedoes
- Air defense systems
- Radars and sonars
- Battlefield communications

A sample of customers includes NASA (National Aeronautics and Space Administration), the DOT (Department of Transportation), and the DoD (Department of Defense). The readers will note that these three are substantial agencies and all are part of the federal government. Despite this potential limitation, it is believed that the lessons learned, in the main, apply to more or less any customer set. So it is hoped that the readers will consider how to use these lessons in their work in systems engineering and related fields. To the extent that this is the case, feedback to the author is welcomed, at heisner@rcn.com. It is also hoped that some of the readers, when they are inclined to do so, will document their own sets of lessons learned.

Howard Eisner
Bethesda, Maryland

About the Author

Howard Eisner spent 30 years in industry and 24 years in academia. In the former, he was a working engineer, manager, executive (at ORI, Inc. and the Atlantic Research Corporation), and president of two high-tech companies (Intercon Systems and the Atlantic Research Services Company). In academia, he was a professor of engineering management and a distinguished research professor in the Engineering School at the George Washington University (GWU). At GWU he taught courses in systems engineering, technical enterprises, project management, modulation, and noise and information theory.

He has written nine books that relate to engineering, systems, and management. He has also given lectures, tutorials, and colloquia to professional societies (such as INCOSE – International Council on Systems Engineering), government agencies (such as the DoD, NASA, and the DOT), and other groups (such as the Osher Lifelong Learning Institute (OLLI)).

In 1994, he was given the outstanding achievement award from the GWU Engineering Alumni.

Dr. Eisner is a Life Fellow of the IEEE (Institute of Electrical and Electronics Engineers) and a fellow of INCOSE and the New York Academy of Sciences. He is a member of Tau Beta Pi, Eta Kappa Nu, Sigma Xi, Omega Rho, and various research/honor societies. He received a Bachelor's degree (BEE) from the City College of New York (1957), a Master of Science degree in Electrical Engineering from Columbia University (1958), and a Doctor of Science degree from the George Washington University (1966).

Since 2013, he has served as professor emeritus of engineering management and distinguished research professor at the George Washington University. As such, he has continued to explore advanced topics in engineering, systems, and management.

Other Books by the Author

Computer-Aided Systems Engineering
Reengineering Yourself and Your Company
Managing Complex Systems – Thinking Outside the Box
Essentials of Project and Systems Engineering Management
Systems Engineering – Building Successful Systems
Topics in Systems
Thinking – A Guide to Systems Engineering Problem-Solving
Systems Architecting – Methods and Examples

Technical

1

1. WHEN AND WHERE POSSIBLE, GO BACK TO FUNDAMENTALS (*)

So what, indeed, are the fundamentals? In a nutshell, they're basic physics and engineering. I would like to illustrate this with a few stories.

Case One

I was having dinner with my son and his two twin sons, my grandchildren. We were exploring entrepreneurship, and what it might take to become one. To make a point, I suggested that there might be a huge market for a device that un-cooks a steak (or the like) when it is well-done rather than rare, as the customer requested. After all, with such a device, restaurants could save huge amounts of money.

"That's a great idea, grandpa", they both agreed. "Let's build such a device".

I encouraged this whole adventure, and they went off with a happy assignment – to figure out how to build a steak un-cooker. Their enthusiasm was almost boundless. If successful, they would become super entrepreneurs – at age 20.

A few days later I got a phone call from one of these grandchildren.

"Grandpa, I have bad news", he said

I replied, "Don't keep me in suspense. What is it?"

"I'm afraid that we can't build a steak un-cooker. It violates the Second Law of Thermodynamics! We discovered that by going back to our High School physics class notes and our textbook".

"Terrific", I said. "You did what needed to be done. You both went back to some fundamentals".

And so the tale ended.

Case Two

Another story goes like this. It was suggested by a friend of mine that I visit with an ex-Israeli who lived in Queens, New York. So when I was spending some time with my brother in the Big City, I went with my sister-in-law to see this ex-Israeli. He showed me an "invention" of his, a prism that took in light and produced the colors of the spectrum as an output.

"There's more energy coming out than the energy going in", he said, spinning the prism around in his hand. "How about we develop this together, and make a fortune", he suggested. With that, he thrust a paper in front of me and urged me to sign it. It was a partnership agreement that, presumably, would get us up and running.

"More energy out than the input energy", I thought. "That simply cannot be. That's more than 100 percent efficiency, and violates a basic Law of Physics".

My sister-in-law, also an ex-Israeli, spoke to him in Hebrew and then to me in English.

"This is a scam", she said, under her breath. "Don't fall for it. He just wants an open pipeline to your money".

I had come to the same conclusion, more-or-less at the same time.

"Sorry", I said, "but I'm going to have to decline your offer, and we have to leave now". And with that, we left his premises and drove back to my brother and sister-in-law's apartment.

There are times, I thought, when one just needs to go back to fundamentals, which also might help in avoiding a poorly disguised scam.

Case Three

I was a young engineer in my 20s and was watching and listening to a senior engineer (actually a physicist) explain his thinking in solving a difficult problem. He did so with grace and a complete command of physics. He set forth a very convincing argument as to how he derived a certain formula and what the "answer" was very likely to look like. I watched and listened in great awe.

"This is what research is all about", I thought, considering his "heuristic" as both correct and masterful.

So there we have it. Three cases that help to illustrate the premise. They're all different, but they demonstrate the point. Stay with the fundamentals, and make sure to be careful as you do so. In particular, listen and don't sign any partnership papers.

2. SERIOUSLY EXPLORE ALTERNATIVES, EVEN IF TIME IS SHORT

This admonition is one of the author's favorites and receives a fully intentioned asterisk [1]. A key issue for the system architect is to take the time to define and evaluate alternatives. The "analysis of alternatives" (AoA) was suggested by the Department of Defense (DoD) as an important part of building new systems. It's not clear as to when and if the DoD will enforce this suggestion; it would be surprising if they did not.

The Department of Homeland Security (DHS) has also explored the notion of analysis of alternatives and has documented their approach [2]. Some of the features of this approach are delineated below:

- First, the alternatives need to be defined
- Then one identifies operational scenarios and concepts of operations (CONOPS)
- This is followed by setting forth effectiveness measures for the alternatives
- Which leads to estimates of cost
- One then plots the values of cost and effectiveness on a graph
- Then this graph is analyzed, in detail

So we see above a basic cost-effectiveness approach. The alternative that is most cost-effective is usually selected unless there are other over-riding factors and influences in play.

Examples of problem areas that are subject to the definition of alternatives include:

1. Buying an automobile,
2. Buying a house, and
3. Buying a computer.

But first we look at a short tale from the world of military communications. Some years ago, I had the pleasure of working on a system known as Consolidated Space Operations Center (CSOC) on a sub-contract for the Air Force. We had won the base contract and moved on to bid on the follow-on effort. In our proposal, we were faced with the matter of what our approach should be – frequency division multiplex (FDM) or time division multiplex (TDM). During the base year contract, we took the position that FDM was the preferred approach. This decision carried the day, and we therefore bid an FDM approach for the second phase contract. As it turned out, a competitive bid came in, taking a TDM approach. That competitor won the competition. As best I can remember, going with TDM was in line with a trend toward digital communications and its inherent compatibility with the computer and various off-the-shelf hardware and software. So we learned a lesson that day: in retrospect, it was conjectured that we should have submitted two bids – one for FDM and the other for TDM. So the wrinkle in the analysis of alternatives sometimes can be not "either-or", but "and". Sometimes it's possible to bid more than one alternative, rather than just one.

While we are on the topic of bidding on contracts, we recognize that in such a scenario where time is usually short, there's a lot of pressure to come up with an answer. So participants argue that there's not enough time to look at more than one singular alternative. This, of course, is not an AoA case since there is only one selection. This author believes that this is generally wrong-headed and that just about all situations call for a legitimate AoA. If one does not do this, then a price is usually paid down the road.

We turn our attention now to the DoD and what their approach might be and how they look at the overall issue [3]. The stated objectives of an AoA, as represented by the Air Force, are:

1. Refine alternatives
2. Refine criteria
3. Refine evaluation factors
4. Work to gain consensus
5. Reduce uncertainty
6. Choose an alternative

We note the following regarding these objectives. First, there is no mention of cost or effectiveness, explicitly. This, in itself, is a bit surprising. The omission of a cost analysis is a most serious matter. Second, the consensus item shows interest in and emphasis on the matter of how decisions are formulated and made. Third, there is the objective of reducing uncertainty. This can be a long and difficult item, with little guidance as to how to do that. The bottom line appears to be that the DHS approach and that of the DoD, as defined here by the Air Force, aren't quite the same. Indeed, the differences are significant. In this comparison, the Air Force approach would appear to be in need of a second draft.

We end this discussion by simply saying that the AoA is the most important notion that should be followed, and refined, for all large-scale systems. Leaping to an intuitive conclusion with only one alternative may well be a mode of operation, but it is not recommended. Take the time to broaden your consideration to at least two alternatives. In the case of architecting systems, there is good and sufficient reason to extend your consideration to three alternatives. See the presentation on systems architecting in this book.

References and Recommended Reading

1. If given an asterisk, it is in the author's "top ten".
2. "Analysis of Alternatives (AoA) Methodologies: Considerations for DHS Acquisition Analysis", Version 3.0, 22 January 2014.
3. "An AoA Handbook – A Practical Guide to the AoA", Office of Aerospace Studies, Kirtland AFB, NW 87117 – 5522, 6 July 2016.

3. EMBRACE PROTOTYPING AS WELL AS MODELING AND SIMULATION (M & S)

Over the years we have seen a stronger move toward prototyping which has had a significant impact on systems engineering. Where this effect started up is debatable, but we can see thrusts in the business world mainly in the form of design. Prototypes can have quite positive effects on a project or system development such as:

a. Reducing the time needed to produce results
b. Costing less, in the long run
c. Moving from "thinking" to "instantiating"
d. Showing the customer(s) something concrete to react to

The latter often leads to customer interjection such as:

"I like it!" or "I don't like it but can you change to something like …?"

In the area of modeling and simulation (M & S), we see continual investment in software. And the reason – to find out how the system should be architected and carry out extensive tradeoff studies that exhibit system and subsystem behavior.

Generally speaking, there are two broad types of M & S systems – event-driven and time sliced. For the latter, we are interested in what is happening at every time slice and standard computer time slices that are of course synchronous help us in this regard. For the event-driven systems, we often wait for "important events which are out of time-step". An example is the appearance of a satellite in and out of view of a ground station as it circles the earth.

This author has attempted to encourage the growth of the M & S field, starting with his book on Computer-Aided Systems Engineering [1], which delineates a substantial list of useful software in several categories, which includes:

- GPSS
- SIMAN
- SIMSCRIPT
- SLAM
- DYNAMO
- GASP
- CSSL

Other categories of M & S-related software that come to mind are:

a. Design tools
b. Alternatives and preference evaluators
c. AI related

A final word on the matter of M & S. Some years ago, this country undertook the challenge of the SDI, or the Strategic Defense Initiative. It was around the year 1990, and I was charged with the responsibility (in my company) for finding out how to participate in this very important program within the Department of Defense. I looked around within the company to see what had the best chance of succeeding in this regard. I looked at the technology that we had built for NASA and concluded that we had something that might apply to the SDI program. This was a piece of software that calculated orbital mechanics and position location on the face of the earth. This might apply to locating a missile that was launched against us from foreign soil (the so-called boost phase of an enemy missile).

I contacted the person in our company who had built this position location algorithm and asked for a demonstration of that software. He was more than happy to oblige since he saw a possible new use for his software, which might indeed be called M & S software, put to a second purpose.

Soon enough we were both driving down the road to Cherry Hill, New Jersey, to demonstrate the software to a potential customer. The basic idea was that the software was capable of simulating the SDI scenario such that we could compute the kill probabilities of enemy missiles during its various stages of flight. This customer was impressed and recommended a contract to explore the SDI scenarios in detail. Soon we were under contract, which was quite a positive experience for both our company as well as that customer. In a relatively short period of time, we were showing the SDI program office what we were able to do to simulate the overall set of SDI scenarios. It was a real breakthrough for the company as well as the customer.

Another experience of this author, in relation to M & S activities, is worth noting. This was the Climatic Impact Assessment Program (CIAP), a two-year study sponsored by the Department of Transportation (DOT). The author's company had a contract with the DOT to support the program, the essence of which was to build a series of models that fit together and that also dealt with the following pieces:

a. A fleet of SST aircraft that emitted various effluents as they flew,
b. The chemistry between these effluents and the atmosphere,
c. The consequent changes in the atmosphere, and
d. The possible increase in cancer, from the above.

The above four components were constructed as a serial model, and it took some two years to do so and then be able to do end-to-end calculations.

Finally, we see several books on M & S systems that will keep any reader up-to-date on this important subject, such as references [2], [3] and [4].

References and Recommended Reading

1. Eisner, H., "Computer-Aided Systems Engineering", Prentice-Hall, 1988.
2. Sokoloski, J., "Principles of Modeling and Simulation", Wiley, 2009.
3. Sokoloski, J., and C. Banks, "Modeling and Simulation Fundamentals", Wiley, 2010.
4. Zeigler, B., "Theory of Modeling and Simulation", Elsevier, 2000.

4. COST-EFFECTIVENESS IS STILL THE PREFERRED APPROACH

The U.S. government, especially the DoD, is looking for:

a. Cost-effective solutions,
b. Cost-effective systems, and
c. Cost-effective architectures.

The intuitive meaning is clear – the more cost-effective, the better. And we can compare two systems based upon their costs and measures of effectiveness.

But we need to have solid approaches to measure both costs and effectiveness, the latter being more complicated.

Let C_1 = the cost of system 1, C_2 = the cost of system 2, E_1 = the effectiveness of system 1, and E_2 = the effectiveness of system 2. If $C_1 < C_2$ and $E_1 > E_2$, we can assert that system 1 is more cost-effective than system 2. Other constructions, such as $C_1 > C_2$ and $E_1 > E_2$, leave us with some uncertainty. System 1 is more effective than system 2, but costs more. Are we willing to pay the extra price (cost) in order to achieve higher effectiveness? We have to look at additional details before we can decide on some answer.

This author supports the basic ideas of cost-effectiveness analysis in order to choose approaches and systems from one another. As suggested above, other groups and agencies appear likewise to support this approach. If we have more than two alternatives, more combinations need to be taken into account, but the basic ideas remain the same.

The notion of cost estimation, of course, is part and parcel of this cost-effectiveness approach. As easy as this approach sounds in principle, significant steps have been taken to support this type of estimation [1, 2]. On the effectiveness side, we conjure up a series of MOEs to generally establish effectiveness measures. These MOEs, by way of example, can be estimated as follows for various types of systems.

Typical MOEs for Communication Systems

- Detection probability
- False alarm rate
- Signal-to-noise ratio
- Availability
- Grade of service
- Speed of service
- Bit error rate (BER)

Selected MOEs for Transportation Systems

- Trip time
- Passenger capacity
- Freight capacity
- Speed
- Connectivity
- Capacity to demand ratio

TABLE 4.1 Weighted (Modified) MOEs Example

ORIGINAL MOES	WEIGHTS (W)	MOE RATING (R)	MODIFIED MOE (W R)
Trip time			
Pax capacity			
Freight capacity			
Maintainability			
Reliability			

Limited MOEs for Air Defense Systems

- Probability of target detection
- False alarm probability
- Target kill probability
- Percent of targets detected
- Number of targets in track-while-scan mode

Weighting Factors

If the various MOEs have different levels of importance, we generally weigh them to obtain modified MOEs. These modified values become the new MOEs for comparative analysis. Such a situation is illustrated in Table 4.1.

There are other ways to approach measuring effectiveness. These are suggested, for example, in Blanchard and Fabrycky's classic text [3]. In that treatise, the authors define several orders of measure, including performance, availability, dependability, producability, sustainability, and others. This is still an open question, but the system analyst, in search of the best design, needs to look at the overall matter of MOEs in some detail.

References and Recommended Reading

1. "NASA Cost Estimation Handbook", NASA Cost Analysis Division, NASA Headquarters, Washington, DC
2. "Parametric Estimation Handbook", International Society of Parametric Analysts, Fourth Edition, April 2008
3. Blanchard, B., and W. Fabrycky, "Systems Engineering and Analysis", Prentice-Hall, 2011.

5. DO NOT ATTEMPT TO INTEGRATE ALL STOVEPIPES

It is well known that there are many "stovepipes" out there, a stovepipe being a system with more-or-less a single focus or function. There is also a great temptation, on the part of management, to try to integrate all stovepipes under a common umbrella system. I have seen more than one directive from an agency head that basically mandates complete integration of the agency stovepipes. In addition, I've been a first-hand witness to an attempt to carry out this form of complete integration.

With respect to the latter situation, I was on an advisory panel for the Navy, and this particular group was trying to integrate six stovepipes to form one overarching system. They had a serious budget, but were running out of time and dollars. Finally, the program to integrate was terminated, and all the contractors and sub-contractors went home, so to speak. I thought long and hard about what had happened and realized that the stovepipes were by no means easily integrated. It was mostly a matter of software, meaning that the structure and languages for these stovepipes were widely disparate. How do you integrate software written in six different languages, supported by several different databases?

Upon further examination, I developed a means by which an "integration index" could be set forth that would be a measure of how hard it might be to integrate two or more stovepipes. The idea was to find a way to help management cope with this not very well understood situation. The essence of that construction is presented below.

The suggestion here is that we accept the viewpoint and ground rule that says:

* Integrate stovepipes only when it is provably cost-effective to do so.

This means that we simply do not try to integrate when it will cost too much and/or it will take too much time, or the overall expenditure of resources will be too high. It also means that we need to brush up on how to do a cost-effectiveness analysis. More about that under another topic in this book.

This matter needs to be studied in detail, for many types of systems, to be more precise about when and how to deal with the "stovepipe issue". How much money and time has been wasted going to the default perspective of issuing a directive that calls for the integration of all stovepipes? My experience and study say that this is a non-trivial problem that calls out for a systematic and comprehensive investigation. Management folks in both industry and government certainly deserve such.

So the "integration" issue remains on the table and is one of the more important ones for the systems engineering community. Here are some of the factors to take into consideration when dealing with when and where to integrate stovepipes:

- What is the ultimate compatibility of the software for the stovepipes?
- What role does size play?
- Are the stovepipes completely mature, or is more work to be done on them?
- What do we actually gain by various integrated (vs. stand-alone) configurations?
- What is the appropriate sequencing if we decide to integrate the stovepipes?
- What are our best estimates of cost for various combinations of stovepipe integration?
- Is there any natural affinity between stovepipes, including overlapping functions?

Another thought with respect to this overall topic has to do with a priori design. By this we mean that it may be possible to plan, in advance, for a future integration activity. If management takes the time, various forms of integration may be envisioned for the future. This is perhaps more than one can expect, given the difficulties faced by program and project managers. These difficulties also have to do with conditions imposed by the procurement regulations.

We complete this discussion by looking briefly at what the INCOSE systems engineering handbook [1] has to say, especially about the integration process. This process is defined with the following purpose: "to synthesize a set of system elements into a realized system … that satisfies system requirements, architecture, and design". Topics discussed in some detail under this topic include:

- The concept of aggregate
- Integration by level of system
- Integration strategy and approaches

The handbook is silent specifically on the matter of "stovepipes" and their integration. This handbook is also brilliantly conceived and written. One might, however, find a few pieces missing, from time to time.

Reference and Recommended Reading

1. D. Walden, G. Roedler, K. Forsberg, R. Douglas Hamelin, and T. Shortell, (eds.), "Systems Engineering Handbook", Fourth Edition, INCOSE, John Wiley, 2015.

6. A BIT AT A TIME OR ALL AT ONCE

There are times when we're looking at doing things incrementally or all at once. They seem to be opposites, and there seem to be natural tendencies to go in one direction or another. Let's look at some of this issue here. Let's see if there's some kind of lesson to be learned in this domain.

In the world of software development, we appear to favor incremental development, at least as a descriptor. One step at time, one increment at a time, distinguished from a "grand design". This is very close to but apparently not exactly the same as evolutionary design [1].

Having said that, we certainly can call an evolutionary approach the opposite of a revolutionary approach. And there is at least one distinct arena in which the latter seems to be preferred. This is in the context of reengineering the corporation, as per the conception of Hammer and Champy [2]. In their blockbuster book in 1993, these authors lay out their approach of how to reengineer the corporation. It was a compelling story, at least as reflected in the super-sales of their first book. And their approach was definitively "revolutionary". After all, they call their treatise "a manifesto for business revolution". Here are some of the aspects of their thinking in this matter [2].

- It's all about processes; we must reengineer faulty and outdated processes
- It's therefore not about people, although people play a key role in carrying out the reengineering of these processes
- In terms of the people side of the equation, thinking needs to shift generally from deductive to inductive
- The overall process is called reengineering and is revolutionary, not evolutionary

So if we're talking about an overall enterprise, the corporation if you will, it needs to be fixed by changing (reengineering) the key internal processes. If we have the right processes, they need to be examined and changed. One direct example of the latter is in systems engineering. We generally have the right technical processes (in Mil Std 15288). Systems architecting, for example, is the right process. However, we're not doing it correctly, according to this author. So the "fixing" involves doing it a bit at a time and on the process of architecting in this example. Throwing out Mil Std 15288 would constitute a revolutionary approach, which we appear to not be doing.

So when do we do the "little bit at a time" and when do we do the "revolutionary" or "all at once"? The answer seems to be – it depends. It depends upon the nature of the problem and the nature of the processes in question.

Let's try to clarify and place in a larger framework the nature of and difference between inductive and deductive thinking. Here's what Hammer and Champy say about the need for more inductive thinking. It's about first recognizing a "powerful solution" and then looking for the problems it might solve, including problems the company doesn't even know it has.

Finally, is there a lesson learned in this arena? Possibly it has to do with the central theme of processes in relation to the "one little step at a time" vs. the revolutionary approach. Perhaps it's something like:

- Try one small step at a time when you're convinced you've got the right process
- Try the revolutionary approach when you believe you might have the wrong process

If we look at the latter suggestion, we seem to think we have the right process(es) in the systems engineering world (i.e., Mil Std 15288). No need for a revolution.

This needs to be pondered for a while. Is there a better or different lesson learned here? What might that be?

References and Recommended Reading

1. Eisner, H., "Essentials of Project and Systems Engineering Management", John Wiley, 2008.
2. Hammer, M., and J. Champy, "Reengineering the Corporation", Harper Business, 1993.

7. GROWTH BY ACQUISITION

During the approximately 50 years in the field of systems engineering, I experienced two purchases (we sold the company twice) and acquired two companies. Here are possible lessons learned from these various events.

The most significant of the above was when we were about an $80 million company and began to take the matter of growth through acquisition quite seriously. We prepared some material that described the company and its features (e.g., LOBs, revenue, profitability, etc.) and made some special trips to companies that were good acquisition targets. In parallel with that, we voted internally on the matter of acquisition as a strategy. The internal vote of our top four executives was three yeses and one no. It was the CFO who wanted to run a bit longer without acquisition. Finally, he changed his mind when we drew up more serious plans, including financials.

This overall process continued for about a year and we learned a lot just by "walking" and "talking". There were all kinds of companies out there for acquisition, but relatively few for our serious consideration, as we had set the bar rather high.

Finally, we ran into two companies that looked good to us. These were Intercon Systems and Calculon. Not too big, not too small, and workable from a financial point of view, which were key criteria for acquisition. The other criteria had to do with the lines of business and overall business sustainability.

Over the years, we had wanted to enter the world of information security but without much direct success. That became a more important criterion, and Intercon Systems had several contracts with a security agency. Put a heavy-duty check next to Intercon. Its headquarters was in Cerritos, CA, and it had about six small offices around the country. These offices were in the San Francisco, Santa Clara, Thousand Oaks, Seattle, and Huntsville areas. Typically they had about a dozen people, just right to serve a special customer and to use as a base for further growth. All of them had very good prospects for contract re-competes. Each had a strong first line manager who would fit very well into the job without appearing to be a problematic person if acquired.

While on the subject of people, we took the time to examine each and every person that we considered important as part of an acquisition. Would they be an asset or a liability going forward? Did they fit into our company or were they going to be a problem child? This "once-over" was an important part of the process, but I don't recall that any of our negative results serving as a deal-breaker. But keep in mind that it might have been otherwise as time went on and new data observed.

The Intercon acquisition turned out to be especially important in terms of this treatise, since after a short while, I took on the position of president of that company when the then president confirmed his desire to retire with his new nest-egg. It was a challenge for me, an East-Coaster who had to make several lifestyle changes with the new position. Here are some changes that I experienced, considered (not all) in this connection.

a. Change of location

The headquarters of Intercon in Cerritos was a problem for me since I had been located in Rockville for many years. The problem was solved initially by my taking a furnished apartment in Seal Beach which I occupied, more-or-less, every other week. But it did mean that I was on an airplane to the West Coast bi-weekly. As it turned out, I fit that into my life with relative ease since I was between wives at that time.

b. New names and faces

The new company had many new "names and faces" that had to be learned, and quickly. When on the West Coast, I arrived with a standing early on Monday morning with key vice presidents. Quick status review and discussion of ups and downs as well as business development activities.

c. Each office key player

Each of the company's eight offices had a key player and a customer. Guess what? I needed to visit with each as a first order of business. What better to do than to see these people? I can't think of any. Can't beat direct contact and time spent with these people as a "lesson learned". Remember, once the acquisition is a done deal, these office directors are your people and your responsibility, and they may be the reason you succeed or fail to do so.

d. Business base and growth for each office

Finally, I felt that I had to make an initial judgment as to what the business base represented in terms of short- and long-term growth, and that's exactly what I did. That involved both strategy and tactics and reaching some type of mutual understanding with the local manager and his boss. Lots of different results, but an important thing to have done.

e. Fit within original acquisition company

Finally, I had to check with my boss, the president of the original acquirer, which was ORI, Inc. Other members of that company's executive committee also had to be kept in the loop. All around communication was an important ingredient in making the new and much larger company work efficiently and smoothly.

As we go through what one might consider after the acquisition is completed, I'm reminded of a story that went around the industry regarding the aftermath of an acquisition. After a year or so of "due diligence", it was discovered that the acquired company could not complete a particular contract within schedule and budget. The contract was a fixed price and the customer insisted that they keep working until all the deliverables were satisfied. This amounted to a massive overrun, and a significant real loss for the acquirer. This, of course, is a path to be avoided on the road to growth by acquisition. When reviewing the field for possible acquisitions, it's crucial to do a deep due diligence that includes what might be called a risk assessment. Keep your best engineers and lawyers busy, before the acquisition, looking for problem areas. One of these, of course, is the fixed price contract that's in trouble. You're better off fixing this problem before the acquisition rather than trying to do so after the deed is done. Lesson learned? One ounce of prevention, in this domain, is worth …

8. THE CONTRACT

As systems engineers, we've come a long way on the contract side. When last I looked, there was considerable participation. And this required a fair amount of preparation as well as a continuing check:

- Are we doing what the contract requires, as written in black and white?

Many moons ago, when my company was small and I was starting out as a young project manager (PM), I got a call from our contracts person. He wanted me to attend a session on the recent contract we won with the Goddard Space Flight Center. I was the PM for that contract and we needed to negotiate a signing of that contract.

"What do I need to do"? I asked.

His answer was simple. "Just be at my office at 1 pm. I'll drive the two of us to Goddard".

That was just fine with me as I discovered the issue was simply to arrive at a rate (fee) for the contract. I did my best, and we wound up with an 8% rate, as I remember the meeting.

The issue is much more complicated these days. The technical persons are more numerous as the contracts have gotten more complicated. And we see cost plus as well as firm fixed price and Time and Material (T & M) contract types. More complex – more of a role on the technical side of the house, including vice presidents.

In today's world, the systems engineer needs to make sure we are indeed fulfilling all of the details of the "deliverables" clause, and on time. It's often the case that the lead systems engineer writes that deliverable and knows most about it. Going beyond mere delivery, there's the question of "evaluation criteria". That means, how do we know that a particular deliverable is acceptable? Sometimes that's easy, as in, have we actually delivered the deliverable?

Sometimes it goes more deeply, as in, does it satisfy the criteria? We need to have the technical folks present for that discussion, and often in the presence of a government representative on the other side of the table (the contract officer).

Bad things can happen if the "technical" systems engineer fails to do his or her job. On a contract that I saw that went along with a company acquisition, due diligence demanded a halt and a deep analysis of that contract. The lead systems engineer failed to flag the problem (over-spent in cost and schedule), so an overrun on a fixed price was the consequence. This led to an overrun on the overall contract of a significant amount.

So the lesson learned can be simply put:

- Systems engineers of today better be prepared to participate in the contract negotiation and also track the clause progress, one at a time, during the life of the contract.

If you can do the above, you're much more likely to be successful as systems engineer and also more likely to eventually be promoted to the next level of management (if that is what you desire). The contract, if you will, is part of the "system", and that is not to be taken for granted or considered to be out of scope for the systems engineer.

9. LESS PAPER PLEASE

Your list of deliverables on that contract would boggle the mind.

If we take a quick look at the Department of Defense Architecture Framework (DoDAF) [1] and the various views that need to be documented, we decide that we'd better start now if we want to fulfill the contract requirements.

E. Rechtin [2] provided us with an applicable heuristic:

"amid a wash of paper, a small number of documents become critical pivots around which every project's management revolves".

Very good advice, and can I find an example of possible excessive paperwork?

I've been told this is excessive but none-the-less called for under DoDAF [3]:

AV-1 – Overview and Summary Information
AV-2 – Integrated Dictionary
OV-1 – High-Level Operational Graphic Concept
OV-2 – Operational Node Connectivity Description
OV-3 – Operational Information Exchange Matrix

SV-1 – System Interface Description
TV-1 – Technical Architecture Profile

Here's the next layer of required deliverables pertaining to the "systems" view:

SV-2 – Systems Communications Description
SV-3 – Systems Matrix
SV-4 – System Functionality Description
SV-5 – Operational Activity to System Function Traceability Matrix
SV-6 – System Information Exchange Matrix
SV-7 – System Performance Parameters Matrix
SV-8 – System Evolution Description
SV-9 – System Technology Forecast
SV-10a – System Rues Model
SV-10b – Systems State Transition Description
SV-10c – System Event/Trace Description
SV-11 – Physical Data Model

And moving on from there to:

The Metamodel
- Conceptual
- Logical
- PES
- IDEAS Foundation Ontology

Viewpoints and models
- All Viewpoint
- Capability Viewpoint
- Data and Information Viewpoint
- Critical
- Project Viewpoint
- Services Viewpoint
- Standards Viewpoint
- Systems Viewpoint
- Models
- Model Categories
- Levels of Architecture
- Architecture Interrogatives
- Architecture Modeling Primitives
- Mapping

That's a lot of paperwork, and a lot of person-hours to produce it.

Going beyond this sample of paperwork, we note that the systems engineering standard 15288 has a standard for Technical Reviews and Audits (IEEE 15288.2). These reviews and audits are listed below and exhibit additional paperwork that might be associated with the 15288 standard.

- Alternative systems review
- System requirements review
- System functional review
- Preliminary design review
- Critical design review
- Test readiness review
- Functional configuration audit
- System verification review
- Production readiness review
- Physical configuration audit
- Software requirements and architecture review
- Software specification review
- Integration readiness review
- Flight readiness review

This is a deep and pervasive problem for those that have competed for and won government contracts. People are assigned to write the deliverables, many of which have no real connection to what is happening on the contract work itself. This area is in need of reform but it seems that there is little forward motion in this regard. But it is also true that spending less, or saving money, is generally not a favorite cause in the military industrial complex that we have come to know (and love) so well.

References and Recommended Reading

1. Department of Defense Architectural Framework (DODAF); see DoDcio.defense.gov.
2. Rechtin, E., "Systems Architecting", Prentice-Hall, 1991.
3. Mil Stad ISO/IEC 15288 and 15288.2, Department of Defense.

Management

2

---

10. DEFINITELY MEASURE, BUT DO NOT OVER-MEASURE

I had an occasion, some years ago, to see a senior manager request (of a junior researcher) a plan for software measurement in a government agency. As I recall, the researcher came back a few weeks later with an extensive list of about three dozen items. The overall scenario had to do with ultimately suggesting that this list be the "standard" for all serious software efforts in that agency.

There are times when one wishes to measure software on a project, but measuring 36 items or such is just too much. Too much time and energy in that case are going into measurement, more than the software development itself. If implemented, it's very likely to sink the project rather than improve the likelihood of success.

There may well be a domain where one wishes to have a laundry list of measurements. I would suggest that such be the case with purely scientific studies. By that I mean science in distinction to engineering. When you're trying to understand the physics of a situation, lots of measurements may be appropriate.

Software Measurement

To help clarify the point with respect to software, we list here some of the measurements that one might wish to employ for a project [1].

- Requirements volatility
- Size of software

- Software complexity
- Progress on software builds
- Computer hardware utilization
- Performance of software builds vs. requirements
- Staffing for each software build
- Software reuse, if any

This list covers quite a few software areas in terms of measurement, but by no means all that one can think of. Possibly that's all one might need or desire. In any case, it's an example of the message of this point.

Overall Project Measurement

The status of all projects in an enterprise needs to be measured at least every month. In this regard, the project manager needs to get up and present the project status to his or her boss and possibly to the vice president in that chain of command. As to what needs to be measured and presented, here's another list that might be a good candidate:

1. Timeline for important milestones vs. original plan
2. Deviations from the original plan in a time dimension, with reasons
3. Cost for current activities vs. original plan
4. Deviations from the original plan in costs, with reasons
5. Plan for future weeks and solving any and all problems
6. Result of any meetings with the customer
7. Personnel changes and reasons
8. New/updated estimates of time and cost to complete

Can you think of any others that you would insist upon hearing about if you're the "boss"?

Significant Parameters

In deciding what to measure, we keep in mind that there's a notion that will help. That has to do with *key performance parameters* (KPPs) and *technical performance parameters* (TPPs). We are urged to define and keep track of these parameters, both directly related to the performance of the system, as this is revealed through design and testing. Here are some parameters that have been used on a variety of projects:

- Radar range
- Reliability and availability
- Signal intensity
- Noise levels
- Signal-to-noise ratio
- Probability of detection
- False alarm probability
- Processing power gains and losses

Looking back at this text, we see three instances in which there are eight measurements. Perhaps this is a number to be used, in general, much like Miller's seven plus or minus two [2]. To try to limit the amount of measurement, perhaps the "constraint" needs to be – you've got at most eight "measurement slots" in each functional area. What are they?

COCOMO I and COCOMO II

A second "cousin" to software measurement can be construed to be software estimation. That's a form of measurement, certainly mastered and provided to all by Barry Boehm [3,4]. The essence of Boehm's construction is the relationship person months (PM) (effort) = $A(size)^B$, where A is defined in terms of effort multipliers and B is related to a set of scale factors (economies and diseconomies of scale). Size is interpreted in terms of source instructions or an equivalent. More specifically, A is computed using 7 or 17 cost drivers and B depends upon 5 factors. COCOMO is a powerful set of estimation software that is used extensively in both industry and government.

References and Recommended Reading

1. Eisner, H., "Essentials of Project and Systems Engineering Management", Third Edition, John Wiley, 2008.
2. Miller, A. "The Magic Number Seven Plus or Minus Two", Psychological Review, **63**(2), 1956.
3. Boehm, B., "Software Engineering Economics", Prentice Hall, 1981.
4. Boehm, B. et al, "Software Cost Estimation with COCOMO II", Prentice Hall, 2000.

11. UNDER PROMISE AND OVER DELIVER

Conventional wisdom these days appears to be to under promise and over deliver. So I've been wondering whether or not this is one of my lessons learned. Here's my non-conclusive answer, which is mostly a fuzzy conjecture.

But before I launch, I asked my son, when he was in his early 50s and was an executive in a substantial company, what was (if he could pinpoint the reason) the basis for his success. He hesitated not a bit and said,

- "I always try to under promise and over deliver".

I registered that with the ultimate duty of a father, so I could bring it up here and talk a bit about it.

Of course it makes a lot of sense, most of it intuitive. Control, if you will, the expectations of your customer, and then you're free to surprise him or her with an outstanding performance.

If I look at several of the project management experiences I've had, I find that I basically tried to keep the customer happy and satisfied with what I was delivering. I did just about all that was needed to achieve this result. And I tried to plan ahead so that the work load was predictable and smooth. Every now and then I was surprised and ambushed; the plan didn't work but mostly due to some new and not very anticipated force.

So did I consistently and deliberately under promise and over deliver? The answer is "no". And would I change my approach now? The answer is still a "no".

A formal study result in this domain [1] appears to support the notion that you get no "extra credit" by under promising and over delivering. That may be counterintuitive, but there it stands. If I look at some of these external experiments and also my natural inclination over the years, I come to this conclusion:

- No under promising and over delivering.

How's that for a lesson learned?

So I try to deeply understand what it is that my customer wants, and also what he or she needs, and go for that end result. I try to find useful and responsive pathways that the customer will appreciate. I try to produce answers that will enhance the long-term relationship between me and my customer.

Over the years, and happily, that's been good enough.

Reference and Recommended Reading

1. Why "Under Promise and Over Deliver" Is Terrible Advice *Inc*. Copy.

12. TRY TO IMPROVE OVERALL SYSTEMS ENGINEERING PRACTICES

Since its inception, INCOSE has sought to improve how systems engineering is carried out both nationally and internationally. This effort is reflected back to its members, and so I remain an individual with a charter to see what I can do along these lines. Looking back on these last 50 years, I can see several areas that can be discussed.

The first area is that of system architecting. As explored in another section of this treatise, significant milestones have been achieved by Eberhardt Rechtin and the DoD. The former set the basis for a deeper understanding of systems architecting and represented a seminal contribution. The latter informed industries (and other parts of the government) how to do systems architecting. Further, this author suggested an approach that is different from that of the DoD. Lesson learned? Keep an eye on new developments anywhere they might appear and bring them into practice as soon as possible.

In previous sections of this treatise, we refer to some of the deficiencies of the DoDAF approach. Rather than rejecting this entire view, a complementary approach is suggested, namely, the Eisner's Architecting Method (EAM) approach. This is a completely different idea that has its roots in the cost-effectiveness analysis of alternatives. This is believed to be a significant improvement that can also be treated as an addition to DoDAF. New perspectives like this almost always help in leading to improvements somewhere down the road.

The second area in terms of systems engineering practices is that of model-based systems engineering (MBSE). This area has gained substantial traction over the years to the point where it is totally accepted by the systems engineering community. So, if you are a practitioner of systems engineering, you very likely need to look into MBSE and how to use this approach.

A third area of no small significance is that of systems engineering itself. The overall field has some 30 elements, but the practitioners have taken the point of view that these elements can be tailored to any and every particular program, depending upon its stated or inferred needs and requirements. This has provided significant flexibility and made the application process more efficient.

Along a similar line of thought, work on systems engineering standards has progressed rather well over the years. A major milestone has been that of Mil Std 15288 [1]. This standard has the following selected technical processes and has been important to the field and its understanding.

  a. Mission analysis
  b. System requirements definition
  c. Architecture definition
  d. Design definition
  e. System analysis
  f. Implementation
  g. Integration

Another area in which significant lessons have been learned has to do with defining and certifications as well as a Masters degree program at various universities. For the latter, we have seen departments shift from industrial engineering to systems engineering, and the latter combined with engineering management curricula. In other words, systems engineering is more than a legitimate area of study within the overall domain of engineering.

Yet another area that has seen major advancements is that of software engineering. This is firmly planted as a subset of systems engineering, and at least two thrusts need to be cited here. One is the entire area of cost estimation, which has been moved forward by Barry Boehm's COCOMO I [2] and COCOMO II [3]. Barry is the leader in this arena, and his approach has been rather spectacular, as has been the research and documentation of the likes of Harlan and Shaw [4], and others [5].

Another area to be cited here has to do with research areas suggested by the Systems Engineering Research Center (SERC) [5]. In November 2019 the following areas were selected from the Research Review:

  a. Mission engineering
  b. Digital engineering transformation
  c. AI and systems engineering
  d. System security engineering

Each significant of the SERC, it is presumed, will suggest research areas that are worthy of further in-depth analysis.

References and Recommended Reading

1. Boehm, B., "Cost Estimation with COCOMO I," Prentice-Hall, 1981.
2. Boehm, B. "Software Cost Estimation with COCOMO II", Prentice-Hall, 2000.
3. R. Taylor, N. Medvidovic and E. Dashofy, "Software Architecture", John Wiley, 2010.
4. Boehm, B., and Rich Turner, "Balancing Agility and Discipline", Addison-Wesley, 2004.
5. Systems Engineering Research Center (SERC), Department of Defense, https://sercuarc.org/.

13. NEGOTIATE

There's a simple solution to many problems in the systems engineering world. Simply put, it's known as the win–win solution. We need nothing more than that in most situations. But getting there may be difficult.

If we go back a few years we run into Cohen's book on negotiation [1] that declares it's a matter of three critical elements:

1. Information,
2. Time, and
3. Power.

These basics are rather fundamental, and by way of example, we are led to another section of this treatise where we point to a situation that requires some consideration in the domain of building systems.

Boehm and his team were building a system that had a response time requirement of one second. The team was not able to meet that requirement. So they analyzed the system in great detail and found that they could indeed meet a requirement of four seconds, but at an increased cost from 30 million to 100 million. This, of course, was a shocking result, but no amount of analysis led to a different and more favorable result. What to do?

After much internal discussion, they decided to try to negotiate an "answer" at the response time of four seconds. The essence of that discussion had to do with the need – that the government really did not "need" a one-second response time, and that four seconds would do just fine. It made all kinds of sense to go down that road and thereby live with the original budget of 30 million for the system. This was not a simple negotiation – but the "story" had to be told in considerable detail. And that's what was done.

If we look at the first sentence above, we see another approach that is simple and straightforward. Look for a win–win solution. What is it that might work for both you and your customer? What is it that's acceptable to both parties?

Although Cohen's treatise goes back some 40 years, this author finds that his key points are relevant today. A "win–win" solution is well known to many and brings the parties to a sense of mutual satisfaction. How to get there? Cohen suggests three steps that will lead to:

a. Building trust,
b. Achieving mutual commitment, and
c. Keeping opposition under control.

Moving ahead from Cohen, we now take a look at what Rob Jolles, a coach and author, suggests on this subject. Jolles uses basketball coaching scenarios to illustrate some of his points. Several of these take note of the fact that some coaches are willing to crash (lose) on the court and pick up a referee "no-no". They know that this will "tilt" the table and lead to stress at the end of the game. So a true win–win consideration has a minimum of ego and a lot of consideration of "the other guy". It's lots of ego suppression here and also lots of willingness to swallow one's pride.

Think about it for a while. When Trump negotiated with China, was it win–win or lose–lose? A non-trivial consideration and distinction. But then again, his style of negotiation was (and is) super non-conventional (and obscure), to say the least.

References and Recommended Reading

1. Cohen, "You Can Negotiate Anything", Bantam Books, 1980.
2. Boehm, B., "Unifying Software Engineering and Systems Engineering", Computer Magazine, March 2000.
3. Jolles, R., "The Fairytale called 'win–win'", Jolles Associates Inc., 11 March 2019.

14. UNDERSTANDING THE ENTERPRISE

The company or enterprise has basic attributes, activities, and features that are well worth spending some time thinking about. A lesson learned in this connection is to try to fit into the enterprise in the best possible way and with minimal dislocation. It means better a priori understanding of what the enterprise is up to and how it does its business. It also means a clearer view of the mission of the enterprise.

Drucker has emphasized these types of issues over the years [1]. What is the company up to? What is it trying to do? What is it doing? What should it be doing? What's its overall strategy? Good answers to these questions can make systems engineering a more integral part of the company's success.

Many students of the enterprise have made the seven S's a way of better connecting the employee/manager to the company. These are [2]:

Strategy. What's the overall success strategy for the company, and how well does this appear to be working?

Structure. Is the company well organized to implement its strategy and carry out its various missions?

Skills. Does the company pay attention to assuring that it has the various skills needed to implement its strategy and tactics?

Systems. Does the company have well-defined systems that carry out its business functions?

Style. Does the company have its own identifiable style?

Staff. Are there specific people who implement the organizational elements and position?

Shared values. Is there a well-documented set of values that can be recalled and pointed to by all employees, and which they all subscribe to?

Monthly Measurements

We recognize, each month, a set of measurements that help to define both what's important in the enterprise and how we appear to be doing. In these monthly measurements, vice presidents report to the president and various middle managers "tell their story" to the vice presidents. We get to see how the various lines-of-business (LOBs) are doing, and we get to ask questions in these areas, depending upon who we are. We also typically suggest changes that we think will make improvements, depending upon who we are.

The Balanced Scorecard

The balanced scorecard is often part of the above in many enterprises. Areas covered in exploring "balance" include such items as [2]:

1. Internal business processes
2. Financial measures
3. Customer interactions
4. Learning and growth in various well-defined areas

Finally, we look for a sense of balance that goes beyond purely financial matters.

References and Recommended Reading

1. Drucker, P., "The Wisdom of Peter Drucker from A to Z", see www.inc.com, 2009.
2. Eisner, H., "Reengineering Yourself and Your Company", Artech House, 2000.

15. THE SYSTEMS APPROACH

Embedded in 50 years of systems engineering study and practice is the systems approach; this author has conceived of and documented this approach [1], which is re-iterated here:

1. **Systematic process**. At this point in time, systems engineering is driven by Mil Std 15288. This, in turn, has been defined by processes and ways of executing these processes [2].
2. **Interoperability**. Emphasis is placed upon the interoperability and compatibility between the subsystems. These subsystems are explicitly defined so as to implement the required system functions.
3. **Analysis of alternatives**. In terms of the systems approach, this implies that we will be looking at alternative architectures. Also, the procedure has been brought under the umbrella of the systems approach although it did not start there.

4. **Iteration**. We use iteration as a positive way to refine the overall system design and carry out implementation.

5. **Slow-die system**. We look for a way to assure that the system will lose capability slowly rather than quickly. This is sometimes called a "slow-die" system and is distinctly part of our systems approach (see also the discussion of resilience as part of item number 35).

6. **Agreed-upon requirements**. Once the system requirements are agreed upon, they are inviolate. However, there is room for negotiating and changing requirements during the development process.

7. **A cost-effective solution**. As part of the systems approach, we carry out a cost-effectiveness analysis and drive toward achieving a cost-effective system.

8. **Sustainability**. Add this to our list of "-ilities" as essential to the systems approach.

9. **Technology and risk**. Be prepared to use advanced technologies in our systems, but with levels of risk that are reasonable and measured. This implies that a formal risk analysis is likely to be part of our systems approach.

10. **Systems thinking**. This is the last element of our well-defined systems approach. It is also the fifth discipline in Senge's conception of the five disciplines [3].

The above list is the author's conception of the systems approach and has been constructed over the years in working on systems as well as systems research activities. A word or two is called for with respect to the last item, namely "systems thinking". According to Senge, this fifth discipline integrates and brings together the other disciplines (mental models, shared vision, personal mastery, and team learning), thus creating the learning organization. It's a holistic approach that sees beyond the pieces, but also sees and understands the relationships and interactions between these pieces.

Systems Thinking

There appears to be quite a lot of attention paid to this last item, above and beyond that of Peter Senge. As noted above, he calls this the "fifth discipline" and pays a lot of attention to it, but he's not the only one. Here's a brief sample of some others:

J. Gharajedaghi [4]. This author connects his approach to managing chaos and complexity. He uses holistic thinking and the design of inquiring systems.

D. Meadows [5]. Meadows approaches this subject as part of the Club of Rome group that relied in part upon "systems modeling", which was derived initially as part of the system dynamics structure.

S. G. Haines [6]. Haines contrasts machine age thinking with systems thinking, which consists of four basic concepts – levels of living systems, laws of natural systems, a systems model, and changing systems that are going through their natural life cycle.

Boardman and Sauser [7]. These authors take a novel and comprehensive approach to systems thinking, dominated by togetherness, engineering dynamics, and complexity.

Michael C. Jackson [8]. Jackson takes his readers through such interesting topics as creative holism, creativity and complexity theory, strategic assumptions, and soft systems methodology.

Peter Checkland [9]. Checkland zeroes in on soft systems thinking and methodology, four key steps, and management thinking.

Monat and Gannon [10]. These authors use concepts, principles, and paradigms to analyze the structural properties of complex systems and their intra- and inter-relationships.

INCOSE Handbook [11]. The Waters Foundation articulates some of the essential features of a systems thinker, including the big picture, increasing understanding, use of successive approximation, and resisting the temptation to come to a quick conclusion.

References and Recommended Reading

1. Eisner, H., "Topics in Systems", Mercury Learning and Information, 2013.
2. Walden, D. et al., "Systems Engineering Handbook", Fourth Edition, INCOSE, John Wiley, 2015.
3. Senge, P., "The Fifth Discipline," Doubleday, 1990.
4. Gharajedaghi, J., "Systems Thinking", Elsevier, 2006.
5. Meadows, D., "Thinking in Systems", Sustainability Institute, 2008.
6. Haines, S. G., "Systems Thinking and Learning", HRD Press, 1998.
7. Boardman and Sauser, "Systems Thinking", CRC Press.
8. Jackson, Michael C., "Systems Thinking – Creative Holism for Managers", John Wiley, 2003.
9. Checkland, Peter, "Systems Thinking, Systems Practice", John Wiley, 1999.
10. Monat, Jamie, and T. Gannon, "Using Systems Thinking to Solve Real World Problems", Systems Engineering Program, Worcester Polytechnic Institute, 2017.
11. "Systems Engineering Handbook", INCOSE, p. 20.

16. INDUSTRY/GOVERNMENT INTERACTION

There are many sources of information that feed into one's ideas about lessons learned – such has been the case with me over the past half century. Perhaps the most dominant, and useful, has been INCOSE (International Council on Systems Engineering). This organization has supported journal articles and conferences at a steady pace for all these years. Interactions between members have also been beneficial to all, and part of the overall package.

Government support has also been extremely useful. A key office for systems engineering within the DoD (Department of Defense) plays an especially important role. This office sponsors the SERC (Systems Engineering Research Center), which is recognized by all three sectors as a significant contributor. A recent conference emphasized the following research areas:

a. Mission engineering
b. Digital engineering and transformation
c. AI/Autonomy and systems engineering
d. System security engineering
e. Systems engineering
f. Human capital development

Universities

Another recipient of support, in general, has been the universities. This sector develops systems engineering curricula and research. The latter comes in several forms including class papers, special assignments, and dissertations. It appears that systems engineering has been growing within academia, especially in the Masters program, but not neglecting doctoral studies. Here is a sample of academic activities in systems engineering which indicates its depth as well as breadth.

- Drexel Institute offers an MS degree in systems engineering as well as graduate certificates in systems design and development, systems engineering, systems engineering analysis, systems engineering-integrated logistics, systems reliability engineering, and peace engineering.

- Embry Riddle has two tracks for a Masters in Systems Engineering, namely, a technical track and an engineering management track.
- UMBC's Masters in systems engineering has a 30-hour non-thesis program with the following core courses:
  - Systems engineering principles
  - System architecture and design
  - Modeling, simulation, and analysis
  - System implementation, integration, and test
  - Systems engineering project
  - Decision and risk analysis
  - Mathematics and MATLAB fundamentals

Selected areas covered by electives in the above program include project management, cybersecurity, advanced architecture, quality engineering, and networks.

INCOSE Certification

INCOSE has stepped into the academic circle, if you wish, through its certification program, with the following sample of ESEP ( ) Exam preparation topics:

- system requirements definition
- architecture definition
- measurement
- model-based systems engineering
- resilience engineering
- modeling and simulation
- lean systems engineering
- system security engineering
- prototyping
- systems science and thinking
- interoperability analysis
- specialty engineering

In addition, special benefits accrue when parts of the systems engineering community collaborate. For example, this occurred when INCOSE and the SERC developed a Resource Directory for the industrial and systems engineering community. This author sees these forms of cross-fertilization as indicators of the growing strength of the systems engineering world.

17. TRADEOFFS

A key element of systems engineering is that of looking at tradeoffs between important variables and parameters. A lesson learned, if you will, is that this activity needs to be carried out early and with determination.

Here are two domains that illustrate tradeoffs – there are, of course, numerous others.

Risk

Three aspects of risk analysis are:

1. Identification of high-risk areas in a system
2. Risk measurement
3. Risk mitigation

In the first of the above, we pay special attention to a single point of failure areas. By a single point of failure, we mean that the overall system is likely to fail when this failure occurs. Both the Challenger and the Columbia (NASA manned space flight) catastrophic mission failures fall in this category.

In the risk measurement arena, we wish to quantify risk, usually in terms of probability measures. This often involves some type of modeling process that leads to a ranking of high-risk areas. These become candidates for the next step with active reductions of risk.

The third category, namely, risk mitigation, requires the design team to make changes in the real system, going beyond simply talking about change. This involves actions that change the system, one way or another.

For high-tech systems in today's IT world, we are tending to pay special attention to technology-risk relationships and tradeoffs. That is, as we include more and more complex technology in our design, we also experience higher and higher risk. But this is not a theorem of design. It may be that in certain situations, higher tech leads to less risk. These areas need to be explored in detail to determine the tradeoffs between technology and risk.

Detection and False Alarm Probabilities

Another strong tradeoff area is that between electronic signals and both detection and false alarm probabilities. The tradeoffs are well-known, but need to be examined in greater detail to determine the "best" set of variables for a particular design. This would apply, for example, to pulse detection for a radar system. The variables in question are five-fold:

1. Pulse height (voltage)
2. Noise

3. Voltage threshold
4. Detection probability
5. False alarm probability

These are but two examples of tradeoffs in systems. Tradeoff analysis drives us toward desired parameters and variables that are critical in today's systems engineering projects, from large to small systems. It is suggested here that tradeoff analysis can be a more prominent part of systems engineering. When engaged for large-scale systems, changes are effected generally in the measures of effectiveness for these systems. In this connection, we refer back to the citations of some of these measures, as given below for generic communication and transportation systems:

Communication systems
- Grade of service
- Speed of service
- Detection probability
- False alarm probability
- Signal strength
- Noise power
- Signal-to-noise ratio
- Range

Transportation systems
- Pax system capacity
- Freight system capacity
- Distance/range
- Required power
- Storage capacity
- Speed/acceleration
- Capacity-to-demand ratio
- Braking distance

So the essence of a tradeoff is: when one variable or MOE is increasing, the others are decreasing, with other factors being held constant. When several variables are "in motion" at the same time, we have, of course, a multi-variate situation. This is what makes the field of systems analysis complicated, requiring many iterations and passes with several data sets. The analyst must therefore approach such problems with a plan and great patience.

18. RESILIENCE

Resilience engineering has come upon the systems engineering scene with a surprising positive force. It appears in the INCOSE systems engineering handbook [1] with the following definition:

"to prepare and plan for, absorb or mitigate, or recover from, or more successfully adapt to actual or potential adverse events".

If we look at the systems approach elements as part of section 42, we see one that is a "second cousin" to resilience, namely, "slow die". One might see the other as a kind of opposite. For "slow die" we start with a system that is all up and slowly degrades in performance. For resilience, we start with a degraded system and consider the ways in which its performance may be increased. In that sense, they're both sides of the same or similar coin.

Resilience is considered to be an emergent system property [2]. Such a system has some features that will anticipate, survive, and recover from just about any disruption of one sort or another. A variety of attributes may be supported when moving into a resilient state [2], such as below:

- a. Capacity
- b. Buffering
- c. Flexibility
- d. Adaptability
- e. Tolerance
- f. Cohesion

The INCOSE Handbook [1] further suggests that the following be typical outputs for resilience engineering:

- a. Preferred system characteristics
- b. Response to specifically related threats
- c. Recovery of functions or service
- d. Recovery time

Suggested activities (between inputs and outputs) include:

- a. Relevant models
- b. MOEs

    c. How to mitigate threats

    d. Impact analyses for each "solution"

Resilience analysis is likely to become more significant as time passes, and threats of various types potentially increase.

References and Recommended Reading

1. "INCOSE Systems Engineering Handbook", Fourth Edition, INCOSE, John Wiley, 2015.
2. Woods, D., E. Holignagel, and N. Leveson, "Resilience Engineering: Concepts and Precepts", CRC Press, 2003.

Idea Based

3

19. THEY WERE RIGHT: KISS, SIMPLIFY, AND REDUCE COMPLEXITY

Many experienced systems engineers have argued that we need to systematically "Keep It Simple" and reduce the complexity of our systems. One of them was the master systems engineer, Eberhardt Rechtin [1]. I certainly agree. But we need to figure out how to do that. Here are a few suggestions as to what to consider in tackling this matter:

a. Develop measures of complexity that make sense
b. Use these measures to determine what is too complex, and what is not
c. Do the above as early as possible in the system's life cycle
d. Actually reduce the complexity of the system(s) under consideration (implementation)
e. Take note of what it is that increases complexity, i.e., [2]
   - Size
   - Functionality (how many and type)
   - Number of modes of operations
   - Duty cycle (static vs. dynamic)
   - Real-time operations
   - Parallel vs. serial operation
   - Very high performance
   - Number and type of interfaces
   - Degree of integration
   - Human-machine interaction
   - Non-linear behavior

It would seem that reducing complexity is not a primary concern of the system designers. Perhaps that is true. But it needs to be much further up on the list of things to worry about, and actually do something about.

If a reduction in complexity is achieved, then there is a lot more room for increasing reliability by using redundancy more effectively. Here is a real-world case that hopefully will illustrate the point.

I was working on a weather satellite known as Nimbus. Goddard Space Flight Center was the developer and designed a three-axis stabilized system with solar panels that were rotated so that they faced the sun most of the time. This, of course, provided a critical function (i.e., power supply) for the overall system.

As it turned out, the motor drive to carry out this rotation failed, and soon the overall system failed. This single point of failure mode was not rare; we had heard about this kind of problem before. It is one that persists today – for example, with certain manned missions and loss of life. That's about as serious as it can be. And in terms of redundancy, it would have been possible to switch in a parallel (redundant) motor drive when the first one failed. Reduced complexity could have meant that we could have increased redundancy.

I can remember a post-mission failure review in which I explored the situation on Nimbus with the program manager at Goddard. We both agreed that this potential problem could have been avoided by using redundancy. But the point that he made went something like this – "You suggested a different design for several single-point failure situations. How do I choose, as program manager, which ones to consider seriously? After all, I've got dollar and time constraints that I need to worry about – every day".

Of course I acknowledged that point and went back to my desk wondering how to answer his question. I never did, and the world continued to turn. Now I bring the point back to life in this treatise. Yes, the question is: We agree to reduce complexity and make room for redundancy, but how do we do that?

In one respect, it's rather easy. There are measures of complexity out there, and we can examine them critically. For example, in the software arena, there's the software complexity measure set forth by a leader in this field [3,4], Tom McCabe. In this case, cyclomatic complexity (CC) is given by:

$$CC = E - N + 2$$

where E is the number of edges of the code and N is the number of nodes in the code.

More conventional approaches can be found with respect to software at the McCabe website and also in a variety of books on system reliability, and parts of books that deal with that subject [4].

References and Recommended Reading

1. Rechtin, E., "Systems Architecting", Prentice-Hall, 1991.
2. Eisner, H., " Managing Complex Systems – Thinking Outside the Box", John Wiley, 2005.
3. McCabe, T., "A Complexity Measure", IEEE Transactions on Software Engineering, **2**(4), 308–320, 1976.
4. Eisner, H., "Essentials of Project and Systems Engineering Management", Second Edition, John Wiley, 2008.

20. SEEK A BALANCED SYSTEM SOLUTION; DO NOT TRY TO OPTIMIZE OR ACHIEVE PERFECTION (*)

One of the (serious) mistakes made by systems engineers is to try to optimize and achieve perfection. That approach is almost always wrong-headed. It leads to overruns in time and cost, with no provable optimum performance. The saying is – perfection is the enemy of the good, and it is correct. In systems engineering, one is aware of this problem and issue and tries to find another way.

What is that way? It involves reaching consensus on the part of the team and also some form of agreement from the sponsor and often the system "stakeholders". They are part of the "system", although it might not be obvious that such is the case.

Where is there room for consideration of "balance"?

Within the process of architecting, as represented in this treatise, is an analysis or evaluation step after several alternatives have been defined. This step involves "weighting and rating", in which the weights represent how to look at the various evaluations. The same logic can be used to try to achieve balance in both the "analysis of alternatives" procedure and the "cost-effectiveness" analysis. We definitely want to use these weights to factor into the analysis and be a vehicle for achieving balance in our design as in our final system.

Another way to look for balance is by means of the architecting team as it represents different approaches and solutions. In this scenario, the skill of the project manager is called upon to hear everyone on the team and to know what their perspectives are in a variety of design decisions. In the ideal case, a better, more highly balanced system solution will evolve and be confirmed.

On the matter of different views leading to different weights, we present here a set of data for an aviation project in which this author participated [1]. These data are the weights given to a set of evaluation factors by the commissioners heading up the Aviation Advisory Commission (AAC) some years ago (see Table 20.1). The various and disparate points of view expressed by the commissioners were important aspects of their discussions and final report. That was my point of view (not necessarily shared by all the commissioners). However, this author was pleased to assist in the overall project at that time. To be more specific regarding the Commission, the criteria used by the commissioners were:

1. Social effects
2. Environmental effects
3. Service quality
4. System capacity
5. Human factors
6. International economic effects
7. Investment costs
8. Operating costs

The weights ranged from 5% to 40%. This kind of example illustrates how various people can look at the world, even one that they are all familiar with, and come to different answers and conclusions.

Other References to Balance

This author can readily think of three other references to "balance" that will be cited here.

The first is referred to in the context of model-based systems engineering, as follows [2]:

- "Systems Engineering is a multidisciplinary approach to transform a set of stakeholder needs into a *balanced* system solution that meets those needs".

In the domain of systems architecting, we see the following statement from this author [3]:

- "A preferred architecture is a choice among several architectures that is *balanced*, cost-effective, and most congruent with the stated requirements and what the customer is seeking, as tempered by program and/or system constraints".

TABLE 20.1 Weights vs. Evaluation Criteria from [1] Evaluators (Commissioners) 1–9

CRITERIA	1	2	3	4	5	6	7	8	9	AVERAGE*
Social effects	5	10	15	10	15	5	10	21	8	11.0
Environmental effects	20	40	10	15	20	5	15	8	12	16.1
Service quality	20	10	10	15	20	15	15	19	18	15.8
System capacity	10	10	10	15	20	20	15	15	13	14.2
Human factors	5	10	10	5	5	10	5	1	6	6.3
International economic effects	5	10	5	10	5	15	10	10	13	9.2
Investment costs	15	5	20	15	5	15	15	12	14	12.9
Operating costs	20	5	20	15	10	15	15	14	16	14.4

*Columns do not add up to 100% due to rounding errors.

The third example comes from the systems engineering and specialty engineering section of the Systems Engineering Research Center [4], with a brief explanation of specialty engineering:

- "Specialty Engineering disciplines support product, service and enterprise development by applying crosscutting knowledge to system design decisions, *balancing* total system performance and affordability".

Stakeholders

And while we are exploring balance, we must also recognize the fact that various stakeholders are looking for certain features in our systems. If we can satisfy them, we are then likely to wind up with a balanced system. These people are individually interested in:

   a. Cost
   b. Schedule
   c. Performance
   d. Specialty engineering
   e. Sustainability
   f. Environmental effects
   g. Safety
   h. Security
   i. And a host of others (some fitting under specialty engineering such as RMA and ILS)

References and Recommended Reading

1. Eisner, H., "Computer-Aided Systems Engineering", Prentice Hall, 1988, p. 352.
2. Friedenthal, S., et al, "Systems Engineering Overview", Practical Guide to SysML, 2008.
3. Eisner, H., "Essentials of Project and Systems Engineering Management", Third Edition, John Wiley, 2008, p. 286.
4. SERC, SEBoK (Systems Engineering Body of Knowledge), version 1.9.1, 16 October 2018, Stevens Institute of Technology.

21. UNDERSTAND THE POWER, IMPORTANCE, AND CHALLENGE OF FUNCTIONAL DECOMPOSITION

I had a chance to talk to the head of a serious software company about their performance on a serious and coveted contract. Apparently, they got into trouble with their customer who threatened the cancellation of the contract. I asked him:

> "If you don't mind, are you willing to talk about the problem and what you did to fix it?"
> "I don't mind", he answered, "after all, it's all history now. And possibly this story will be helpful to others".
> "Terrific", I said, "and I'll be very private with names and faces".
> And then he told me "his story".

The story had everything to do with "functional decomposition". It was a large system, and they indeed decomposed the system into functions. But they didn't stop there. The functions were decomposed into sub-functions, and these into sub-sub-functions. They then had many, many of those and set the engineers to work, analyzing and looking at data flow within and between these elements. Soon, with much money and time having been spent, they found themselves behind plan in both dimensions. The customer eventually became aware of the problem and insisted that the company president take a more active role to fix the problem. And that's what happened. The overall approach in "solving" the problem involved the following key steps:

1. Start over and stop the detailed decomposition of the system
2. Decompose to only three levels
3. The three levels became the system name (level one), the major functions (level two), and the sub-functions (level three)
4. Trying to assure minimal overlap or interaction between the functions, as described

It turned out that these three levels were enough, and that was all that was needed to start the process of system architecting.

So we see, in the real world, that simplification was indeed what was necessary, and also that functional decomposition played a critical role to start the design and architecting of the system.

And also, can you provide the reader with an example of functional decomposition?

Possibly the most coherent example is a basic "IT" (information) system for which a simple decomposition would lead to the following top-level functions:

1. Input
2. Output
3. Processing
4. Operating system
5. Applications software
6. Security software
7. DBMS (database management system)
8. Storage
9. Networking
10. Power supply

Not exactly rocket science, and it will do as one approaches the next step of synthesis, or the formulation of alternative architectures. More about that in later subject areas.

And in software, let's take a brief look at the comment from one of our well-known software folks [1]:

> The most difficult design task ... is the decomposition of the whole into a module hierarchy.

Yet another of our software gurus wrote [2]:

> from this process (Wirth's suggestion), one identifies modules of solutions or of data whose further refinement can proceed independently of other work.

Also, this author documented the following [3] with respect to the decomposition of an air defense system:

1. Threat assessment
2. Command
3. Control
4. Communication
5. Detection
6. Guidance
7. Identification
8. Surveillance

9. Tracking
10. Kill assessment

And here we have a comment (a heuristic) from our well-referenced engineer, company president (Aerospace), government executive, and teacher (at USC) [4]:

> "Choosing the appropriate aggregation of functions is critical in the design of systems", and "in partitioning, choose the elements so that they are as independent as possible".

Decomposition may appear to be simple, but it can be quite complex for some types of systems, especially when considering software. But it's essential when one considers the matter of architecting a system. Something to always keep in mind.

References and Recommended Reading

1. Wirth, N., "A Plea for Lean Software", IEEE Computer Magazine, February 1995.
2. Brooks, Jr., F., "The Mythical Man-Month", Addison Wesley, 1995.
3. Eisner, H., "Essentials of Project and Systems Engineering Management", Third Edition, John Wiley, 2008.
4. Rechtin, E., "Systems Architecting", Prentice-Hall, 1991.

22. BREAK THE PROBLEM INTO PIECES USING THE REDUCTIONIST APPROACH WHENEVER POSSIBLE, AND THEN APPLY LATERAL THINKING

There seems to be no lack of suggestions as to what approach to take with respect to large-scale problem-solving. In general, we may consider general methods and specific methods. In terms of the former, I can cite the following:

• Systems analysis
• Analysis of alternatives (AoA)

- Modeling
- Simulation
- System dynamics
- Cost-effectiveness evaluation
- Linear and non-linear programming
- Various forms of thinking (inside and outside the box)

and many more.

On the more specific side of the ledger, we have such approaches as (barely touching the surface):

- The systems approach
- Calculus
- TRIZ
- Use of stories and fables
- Econometrics
- Synectics
- Fourier and Laplace transformations and analysis
- Reductionism

Many authors, including this one, have enumerated "steps" in the problem-solving process. For example, several professors from the University of Virginia have cited these steps [1], which they call "phases of systems analysis".

1. Determine goals
2. Establish criteria for ranking alternatives
3. Formulate alternative solutions
4. Rank the alternatives
5. Iterate
6. Take action

Russ Ackoff has set forth several approaches that appear to work in this important domain [2]:

- The need for managers to be good (if not great) problem-solvers, using his "5 C" model of (1) concern, (2) competence, (3) communication, (4) courage, and (5) creativity.
- Using fables, parables, and art.
- The key steps of (1) defining objectives, (2) conceiving of possible actions, (3) exploring the nature of the environment surrounding the problem, (4) setting forth the relationship between the above, and (5) possible constraints.

Einstein has revealed his approach, namely, to:

1. Visualize as much as possible
2. Use pictures of various types
3. Use your imagination, rather than a set of facts
4. Do not necessarily trust logical thinking

Newton appears to have favored such notions as:

1. Finding truth in simplicity
2. Making bold guesses

Da Vinci also emphasized simplification and relying on experience.

This author, in previous works, has cited approaches and steps, including:

- More than a dozen ways of thinking [3], to include a formal treatment of "thinking outside the box" [4]
- TRIZ [4]
- Reductionism
- The steps of (1) defining the problem with some precision, (2) establishing the key factors and variables, (3) setting forth possible inferences from (2), (4) creating potential solutions, and (5) selecting the best solution.

In terms of personal preference, this author would select two approaches as special. The first is reductionism [3], and the second is "lateral thinking" [5]. The former relies on one's ability to break the problem into pieces, solve each piece, and then put the pieces back together. This often works when the pieces are formulated as conditional probabilities that can be multiplied to yield an answer. An example is a program in the Department of Transportation which was cited previously and is known as the climatic impact assessment program (CIAP) [3]. With respect to lateral thinking, it was devised and named by Edward de Bono and used in various systems engineering problem-solving sessions [5]. So much for a very brief citation of a very complex and well-researched issue.

References and Recommended Reading

1. Gibson, J., W. Scherer, and W. Gibson, "How To Do Systems Analysis", John Wiley, 2007.
2. Ackoff, R., "The Art of Problem Solving", John Wiley, 1978.

3. Eisner, H., "Thinking – A Guide to Systems Engineering Problem-Solving", CRC Press, 2019.
4. Eisner, H., "Managing Complex Systems – Thinking Outside the Box", John Wiley, 2005.
5. de Bono, E., "Lateral Thinking", Harper & Row, 1970; de Bono, E., "Lateral Thinking; Step by Step", Harper Perenniel, February 2015.

23. DEVELOP AND TRY A NEW WAY OF ARCHITECTING

There are times when it's necessary to question conventional wisdom. That's one of my nine suggestions for "thinking outside the box" [1], and it applies directly to the matter of architecting a system. I was digging into the matter of how to execute this critical process (i.e., architecting), and I, of course, ran into DoDAF, which is the Department of Defense Architectural Framework. This framework is based upon views, which are cited as follows [2]:

- An operational view
- A systems view
- A technical view

These are views, but where is architecture? Is it the case that the purveyors of this flawed approach believe that one can, consistently, infer what architecture is from these views?

My evaluation is that, in general, the answer to that question is "no". And so we are left without the confidence that we actually have an architecture in hand. From questioning this piece of conventional wisdom, for me the issue was – what's the next step?

Answer – formulate a reliable method for architecting a system that holds up to scrutiny and provides a consistent logical framework. That is what this section ultimately is all about.

The DoD Procedure for Developing Architecture

As part of our journey here in DoDAF-land, let us look further into the recommended procedure, from the DoD, as to how to build an architecture. See the six steps below [2]:

Step one – articulate the intended use of the architecture
Step two – establish the scope of the architecture

Step three – determine the data needed to support the architecture development

Step four – collect, organize, and store the architectural data

Step five – conduct analyses in support of the architecture objectives

Step six – document results in accordance with decision-maker needs

These steps make very little sense in relation to what they are advertised to be, namely, an architecture development process.

Products for Views

Moving down the road with DoDAF, we find an essential set of "products for views". This is guidance for what constitutes a view, and may be reiterated as:

AV-1 – Overview and Summary Information
AV-2 – Integrated Dictionary
OV-1 – High-Level Operational Graphic Concept
OV-2 – Operational Node Connectivity Description
OV-3 – Operational Information Exchange Matrix
SV-1 – System Interface Description
TV-1 – Technical Architecture Profile

These are the key (essential) products, but where is architecture? Can one infer architecture from these products? The answer, for me, is "no". These products may ultimately be useful, as views of something, but it's not clear as to what that something is. This appears to be a classic case of – if you don't really know what you're doing, double and triple down and dig more deeply by providing more and more detail. Perhaps there is something interesting (and useful) down there.

An Alternative Approach

And so, an alternative approach, set forth by this author, involves the following steps:

1. Functionally decompose the system
2. Formulate design choices (at least three for each function); this is a synthesis step
3. Analyze and evaluate the alternative architectures
4. Decide on a preferred architecture, using cost-effectiveness measures

With respect to these steps, we take note that:

- Functional decomposition is critical; we need to confirm the functions that the system is to carry out,
- The synthesis step *is a description of alternative architectures (!); supercritical,*
- By analyzing alternatives, we are carrying out a recommended DoD procedure, namely, an *analysis of alternatives* (AoA), and
- By step four, we return to a tried and true procedure, namely, an assessment of costs and effectiveness.

I have modestly called this latter approach the EAM (Eisner's Architecting Method) and decided to spend a fair amount of time working with it. This can be considered an alternative to DoDAF or complementary to the same. Take your pick. There's much work to be done in this arena, and we cannot afford to have a method, used and mandated by the DoD and its contractors, rule the day. This is too vital an issue. And, in this author's experience, the DoD is usually not this far off in tackling a difficult problem and issue.

References and Recommended Reading

1. Eisner, H., "Managing Complex Systems – Thinking Outside the Box", John Wiley, 2005.
2. DoDAF, version 2.02, DoD Deputy Chief Information Officer, see dodcio. defense.gov.

24. PLATO AND PROUST

So the article in the August 18 issue of the *Washington Post* [1] had the title "Plato and Proust Can't Save Silicon Valley". The implication appeared to be that we're trying to "fix" Silicon Valley by adding some humanities courses to various STEM approaches. These types of "fixing courses" have been added wherever possible, and it's not doing the job.

What's the job to be done?

Presumably to have STEM people and other technologists behave with a stronger sense of morality as well as an appreciation of the humanities and how they fit into today's technology-driven company.

Does this solution make sense?

The author's article says it's not working.

What does all this have to do with lessons learned in systems engineering? This author tries to bridge the gap by claiming that broadening one's approach to systems engineering is itself a lesson learned. And that is indeed the case. A lesson learned is simply that the systems engineering profession needs to be broader, more inclusive, and more concerned with social issues.

Keep in mind that it all, in many ways, starts with the theory and purpose of the enterprise. Keep in mind that a young and very smart technologist is likely to have this type of conversation with Mark Zuckerberg's recruiter, on behalf of his company (MZR):

MZR – So keep in mind that your job is to come to our company and improve its bottom line in terms of revenues and profits, quarter by quarter.

R (Recruit) – I get that. But what about the company's social responsibility?

MZR – We have other people working on that issue, including me.

R – OK I'd like to take you on your word on this matter, but I'd still like to see just a bit of "proof". Do you have anything?

MZR – You bet. Take a look here at our mission statement. Here it is – "to give people the power to build community and bring the world closer together".

R – That's a terrific vision. Very believable.

MZR – Glad you like it. Took a while and a lot of effort to produce it.

R – I'm not surprised.

MZR – That's usually the case. Repeat it and you'll find even more meaning.

R – I get it, and I'm convinced. … But I'd like to know. Do I have the job?

MZR – You sure do. But keep in mind that you're the STEM person, or the STEAM person, or the technologist. You're not the conscience of the company. I've got that one covered.

R – Thank you for the quick response on that matter.

MZR – You're welcome. Now, you're part of the lifeblood of my organization, Welcome aboard.

R – My pleasure. I'm so happy to join the team.

R – Possibly we can start with the literature pertaining to the balanced scorecard. Do you take that approach, and is it available?

MZR – I'll check on that with my boss.

R – Then we can look at open source papers that have been written in the past five years or so.

MZR – All sounds good. …

MZR – So how do we "fix" the technologists? We show them that the enterprise is truly committed to moral behavior as well as a culture that we can be proud of. And that culture involves constantly iterating to solve problems and working together to provide products to more than 2 billion people. And as you might expect, we're deep in various technologies like artificial intelligence and virtual reality.

R – I'm interested in both.

MZR – As you suggested in your resume.

R – And I'm impressed with your emphasis on diversity. That opens up the likelihood that I'll be meeting and working with all kinds of people.

MZR – That's what it's all about.

R – Boy, I better add this to my list of lessons learned as a systems engineer. It may well help me in my job and also help my company become more socially conscious.

MZR – That's good for all. Let's see how we can all pitch in.

Let's move on and move out and get you to meet our greeting team. This will be the experience of a lifetime, working with a company that is a leader in its field and breaking new boundaries just about every day.

Reference and Recommended Reading

1. Musgrove, P., "Proust and Plato can't save Silicon Valley", Washington Post, Outlook Section, August 18, 2019.

25. TRY TO MASTER NEW TOOLS AND USE THEM AS NEEDED

In this wonderful but at times strange world of ours, new commercial software appears quite often. As a systems engineer, I try to keep abreast of these new developments, and from time to time, get into the fray with a purchase or two.

There appear to be several types of such packages that include at least the following:

a. Decision support
b. Diagramming
c. Languages
d. Expert systems

  e. Statistical applications
  f. Spreadsheets

Here are one or two examples that are on my list for now and into the future.

Decision support. An example of a decision support system is "Expert Choice", which has been available through a professor at GWU since about 1983. He has been a colleague from another department (from mine) and was very active in building and promoting his excellent software. My work (from time to time) with that package found it to be very user friendly and based upon what is known as the Analytical Hierarchical Process (AHP) as studied and documented by Thomas Saaty [1]. Of course, there are many other packages that fit under the category of decision support such as Compterra.

Diagramming. I continue to use "Visio" (from Microsoft), which allows me to use a large number of plain vanilla diagrams and charts which might be called generic. These are extremely pleasing to the eye and simple to manipulate. Here again, a relatively straightforward search shows that there exist many diagramming packages with hundreds of features.

Languages. Python is my system of choice at this time in my life, after having used BASIC and PASCAL during several earlier years. Pascal is currently in the first place, but Python has taken the lead for use by a novice such as this author. Two of my grandsons, as college seniors, use Python.

Idea management. Keep this category in mind and take a hard look at "Bright Idea" and others.

Expert systems. For such systems, one generally makes a series of measurements, feeds them into a built-in inference engine, and obtains results that are supposed to emulate the process of an expert. A popular application field is that of medicine, especially diagnostics. Don't expect a lot of use, but do expect a certain amount of experimentation and fun.

Statistical applications. Looks like "R" is "gaining on the outside" and well worth looking at. This is a derivative of "S" and apparently showing popularity among college students. No charge for its use makes it an extremely desirable choice.

Spreadsheets. My system of choice has been EXCEL over the years, and remains so. It's readily accessible (as part of Microsoft Office) and well documented. I've been able to use EXCEL in a modeling application that involved COCOMO [2], and it worked beautifully – no hiccups or problems. This package has more capability than most people seem to recognize and is the first place to explore when looking for a simple modeling package. Lots of "hidden" functionality in EXCEL that come to light if you really dig.

Keeping track of new tools is itself quite challenging. They seem to be appearing just about every day. And when new ones take root, it's quite satisfying, especially for the systems engineer. After all, we're a group that likes to putter with and admire new technology. Lesson learned – stay connected to the very rich set of tools that are available to us as the years go by.

References and Recommended Reading

1. Saaty, T., "Mathematical Models for Decision Support", Springer, 1988.
2. Boehm, B., "Software Engineering Economics", Prentice Hall, 1981.

26. REAL EAM

One of the most important lessons has been in the area of systems architecting. I've been able to construct an architecting method, the EAM, and tested it for more than two decades. This testing has been done primarily with students over the years, but also in the form of workshops and seminars. This method has four steps [1]:

1. Functional decomposition
2. Synthesis
3. Analysis
4. Selection of preferred architecture

We note especially the significance of the first step. There must be clarity in this step, in that, above all, we need to know the functions that the system is to perform.

Given the functions, the synthesis step looks especially at the ways to instantiate each and every function (and sub-function). This step is considered to be the "heart and soul" of architecting. We note that formulating more than one architecture is an integral part of this step. This is also in consonance with the overall "analysis of alternatives" procedure suggested by the DoD [2].

Next, we look at the evaluations of alternative architectures by placing the alternatives in a cost-effectiveness context. Effectiveness is measured by a weighting and rating procedure, as illustrated in Table 26.1 for five MOEs.

The effectiveness measures for each of the three alternatives are 9.20, 7.08, and 6.18. Independently, the cost estimates for these three alternatives are $2,000, $3,000, and $12,000. The analysts may now select a preferred system

TABLE 26.1 Analysis Step for a Hypothetical Information System

MOES	WEIGHTING (W)	ALTERNATIVE A RATING (A)	W × A	ALTERNATIVE B RATING (B)	W × B	ALTERNATIVE C RATING (C)	W × C
Grade of service	.2	.8	1.6	.8	1.6	.9	1.8
Speed of service	.3	.8	2.1	.8	2.4	.8	2.4
Sustainability	.2	.9	.7	1.4	1.6	.9	1.8
Maintainability	.2	.8	1.6	.7	1.4	.8	1.6
Reliability	.1	.9	1.8	.8	.08	.9	.09
			6.18		7.08		9.20

alternative (architecture) as B, as the best-value approach. This need not be the best answer. Think about the various possibilities as we note here the special features of the EAM approach:

1. The overall process begins with a functional decomposition of the system. In general, all three alternatives have the same functionality but different levels of performance for each. At times, we may experience *function creep* as, for example, from C to B to A. This should be avoided. Unlike the DoDAF [1], the starting place for architecting is not the three views of the system. This point is made in some detail in various parts of the author's book on the subject [2].

2. The synthesis step (second step) directly represents the various architectures, by definition. This shows the value of this step and is appealing as a means of comparison. It also facilitates checking for interoperability by scanning from top to bottom. The analyst tends to appreciate having this amount of comparative information all on one page. This is in consonance with Rechtin's KISS approach. Information regarding sub-functional requirements is available but not explicit in this synthesis step.

3. The next step provides an explicit analysis of the three alternative architectures. This process becomes a cost-effectiveness analysis, well known to the many analytic techniques. We are looking for low-cost solutions where we can find them. On the other hand, the knee-of-the-curve region will hopefully represent a best-value solution. We use the graph as a basis for analysis, pointing us in the right direction. At the high end of the cost-effectiveness graph, we see high costs, but also high effectiveness. There are times when such a system is preferred, especially when looking at military systems of various types. This approach to cost-effectiveness analysis of alternative systems may be said to be congruent with the AoA (analysis of alternatives) process recommended by the Department of Defense [2, 3].

References and Recommended Reading

1. DoDAF, Department of Defense, dodciodefense.gov/library.
2. Eisner, H., "Systems Architecting – Methods and Examples", CRC Press, 2020.

27. WAYS OF THINKING

There are many ways of thinking, including fast and slow thinking as introduced by Kahneman in his classic treatise [1]. If we look back at some of our great thinkers, we see at least five ways of thinking that seem to stand out, and that might well be emulated. These appear to be:

a. Visualization
b. Lateral thinking
c. Hybrid thinking
d. Thinking hats
e. Special point-of-view thinking

Each of these is briefly explored below.

Visualization

There appear to be many modes of visualization, all of which have their special ways of problem-solving. A very simple example is that of seeing beyond a math formula into what that formula represents. It's one thing to write the formula for an ellipse or a circle; it's another to actually "see" the ellipse or the circle. It's one thing to write down the formulae for Maxwell's equations; it's another to be able to visualize the electric and magnetic fields as they change with time and space. Similarly, equations can be set forth for gravity and how it might be represented and visualized, and what that might mean in terms of gravitational theory and black holes.

Visualization generally means "seeing the picture". And as it is said, one picture is worth 1,000 words.

Lateral Thinking

This author has found "lateral thinking" especially useful. When you're digging a hole in a particular place and not making much progress (other than producing lots of earth), it may well lead to some lateral considerations:

• Is my approach sound?
• Am I digging in the right place?

- What did de Bono [2] say about these types of questions?
- Could this form of "sideways" thinking lead to new business areas and breakthroughs not previously considered?

Here's a simple example.

You're a company that builds small radar systems. As part of your strategic planning, you think laterally by expanding upon the variables having to do with radars:

   a. Ground vs. airborne vs. shipboard
   b. Various frequencies
   c. Pulse vs. Doppler vs. chirp

And then you think along another dimension, as per police, army, navy, etc.

Are there new and fruitful areas that you and your team have not yet explored?

This opens your eyes to business areas not previously considered. Your strategic plan has expanded and has some new things to explore. This typically is what happens with lateral thinking. All of a sudden there are some new possibilities. All of a sudden the world of potential has increased. All of a sudden you're taking seriously something you had previously overlooked or bypassed.

Hybrid Thinking [3]

This mode of thinking involves some amount of lateral thinking with some degree of "drilling down". One's intuition plays a large role here – how much of each is the right amount? A radar example might be: define both a harbor radar and an airport radar and dig down to see if either appears to make sense as a potentially new business area. Another example might be how to deal with the complex matter of finding a cure for cancer. Hybrid thinking might involve research in two main areas (i.e., radiation, chemo) and then digging down into both of these areas.

Six Thinking Hats

Yet another way of thinking was set forth by Edward de Bono in his work on Six Thinking Hats [4]. This approach, according to de Bono, has been used extensively to constitute productive teams. Members of these teams wear hats that are "modes of behavior" in the group, not descriptions of the people. Here are the hat colors and the focus of each of them:

The *White* hat deals only with objective fact.
The *Black* hat represents caution and seriousness.
The *Green* hat concerns itself with new ideas and creativity.
The *Yellow* Hat is sunny and optimistic.
The *Red* Hat displays anger and emotionality.
The *Blue* Hat suggests control, including that of influencing the other hats. This is a kind of moderator.

These hats constitute ways of thinking in any type of problem-solving session. One may also think, from a systems engineering perspective, that they can be associated with functional decomposition. This brings de Bono's basic ideas into the world of systems engineering and, by extension, into the world of lessons learned in that domain.

Above all, allowing one's approach to new ways of thinking is likely to yield substantive results. With time and practice, one improves in terms of both processes and products. Hopefully, all of this can be productively derived in the systems engineering world and applied to future problem-solving in that and other domains.

Special Point-of-View Thinking [3]

Emphasis on one of our great thinkers has been provided by Michael Gelb [5] in his deeper look at Da Vinci. Here are some of the thoughts set forth by that author:

a. Embracing paradox, ambiguity, and paradox
b. Quest for learning
c. Refinement of the senses on a continuing basis
d. Learning from experience and mistakes

Many of these should feel familiar to most of us.
Some further special sources of thinking may be cited as [3]

a. Aristotle – be critical and continuously evaluate
b. Newton – the truth is found in simplicity
c. Einstein – imagination and visualization
d. Feynman – prove yourself wrong

    e. Russell – Abandoning one's own reason leads to no end of trouble
    f. Kahneman – we operate under a "fast. Slow" model set of behavior

In the opinion of this author, Peter Drucker has made unique contributions to our overall way of thinking, even though he is thought of as a "management or business" consultant [3, p. 51]. One of the most cogent is his emphasis on systematic innovation through the exploration of opportunities. Some of the sources of new opportunities, he claims, are:

    a. Opportunities are simple and focused (not obscure)
    b. Opportunities are for now, not for the future
    c. Opportunities are there, despite surrounded, at times, by risk
    d. Opportunities tend to start small and grow
    e. Opportunities, as innovations, have a positive effect on society

Tom Kelley has also set forth some new ideas and approaches in his company (IDEO) and book [6]. Kelley has given "names" to contributors in three categories, namely:

    a. Learning personas,
    b. Organizing personas, and
    c. Building personas.

A special area suggested in particular by Luke Williams [7] is that of "disruptive" thinking. An often-cited example of such is that of the pipes and vents on the outside of the Pompidou Center in Paris. The idea, of course, is to out-do your competition with new and novel designs, and one way to think about that is to enter the "disruptive" domain of thinking.

References and Recommended Reading

1. Kahneman, D., "Thinking, Fast and Slow", Farrar, Straus and Giroux, 2011.
2. de Bono, E., "Lateral Thinking", Harper & Row, 1970.
3. Eisner, H., "Thinking – A Guide to Systems Engineering Problem-Solving", CRC Press, 2019.
4. de Bono, E., "Six Thinking Hats", Little, Brown and Company, 1985, p. 198.
5. Gelb, M., "How to Think Like Da Vinci", Dell Publishing, 1998.
6. Kelley, Tom, "The Ten Faces of Innovation", Currency Books, 2006.
7. Williams, Luke, "Disrupt", Pearson Education, 2011.

28. NEW IDEAS TO BE EXPLORED

Most of us get continual pleasure from problem-solving and the exploration of new concepts and ideas. That is, as long as they're either not off the scale or trying to convert science fiction into science non-fiction. So I've included here as a lesson learned some of the challenges I've run into or created for myself in the realm of systems engineering.

General Systems Theory

Every now and then I re-visit Bertalanffy's book on general systems theory [1] and try to bring it a step forward, or put forth a "new" idea or two. Then I see that one or more of my colleagues are a few steps ahead of me and has published a paper "Beyond the Edge". This is all energizing and is part of being a systems engineer. Here are a few sources that suggest an attempt to formulate or expand a general systems theory [2,3,4].

Rapid Computer-Aided Systems of Systems

Another example of trying to articulate new ideas and approaches is that of Rapid Computer-Aided Systems of Systems (RCASSE) [5]. This notion suggested that systems of systems engineering could be enhanced by limiting the scope and better use of the computer over time. The elements of RCASSE, according to this idea, are:

1. Mission engineering
2. Baseline architecting
3. Performance assessment
4. Specialty engineering
5. Interface/compatibility evaluation
6. Software issues/sizing
7. Risk definition/mitigation
8. Scheduling
9. Pre-planned product improvements
10. Life cycle of cost issue assessment

With a list of this sort, we are expressing priorities in systems engineering that will lead to shorter timelines and more penetrating levels of analysis.

One might also interpret this step forward as related to "agile" systems engineering.

New Method of Systems Architecting

Another item on this list is a new method of architecting a system. This item is more fully addressed under lesson learned number 21. This is also discussed in some detail in this author's recent book [6].

National Aviation System (NAS) Model

The above category also includes a contractual effort by this author to build a National Aviation System (NAS) Model. This was a landmark study and activity that set the stage for several more detailed models.

Systems Engineering and Software Engineering

Every now and then this author returns to a set of ten areas that the systems engineer needs to know about software engineering [7]

Emergent Properties of Systems

Affordability

The INCOSE handbook [8] has also highlighted the idea of affordability. It is a "balance" concept and is in search of a "theory". Apparently, it has been addressed by the DoD and is well worth its greater exploration. Possibly there will evolve from such an effort some type of metric that is related to cost-effectiveness measures.

Design to Cost

This is a concept, going back to the original standard for systems engineering, that deals with cost goals for components for the "design, development, production and support" of systems. This author believes that the notion is worthy of returning to today's world of systems and their cost-effectiveness measurement.

References and Recommended Reading

1. Bertalanffy, L. V., "General Systems Theory", George Braziller, 1968.
2. Skyttner, H. G., "General Systems Theory", World Scientific, 2006.
3. Boulding, K., "General Systems Theory – the Skeleton of Science", Management Science, 1956.
4. Klir, G., "Facets of Systems Science", Springer, 2001.
5. Eisner, H., "Managing Complex Systems – Thinking Outside the Box", John Wiley, 2005.
6. Eisner, H. "Essentials of Project and Systems Engineering Management", Third Edition, John Wiley, 2008.
7. Eisner, H., "Systems Architecting – Methods and Examples", CRC Press, 2020.
8. Walden, D., G. Roedler, K. Forsberg, R. D. Hamelin, and T. Shortell, "Systems Engineering Handbook, Resilience", John Wiley, 2015, p. 229.

People Oriented

4

---

29. BUILDING A HIGHLY PRODUCTIVE SYSTEMS ENGINEERING TEAM

One of the most serious aspects of successful systems engineering is that of implementing (project) a team. Each member of this team needs to be carefully chosen and must understand his or her place in the team.

I have been a member of highly functional teams and also in a position of team oversight on more than one large-scale system development projects. Here are a few observations and suggestions regarding the building and workings of a systems engineering team.

The Team Leader

An effective team leader is the number one requirement for a highly productive team. This leader must be technically competent and mature in the domain of the system that is being built. He or she must also have excellent "people" skills and be able to pick every member of the team. Serving as a member of this team should be considered special. Being in the team means that one has a new boss during the period of the team's existence. This new boss, the team leader, has the usual responsibilities of being a boss.

The team leader must also assure that all team members get a chance to speak their minds on everything regarding the system's development. That includes both technical matters and the systems engineering process that is being followed. No member of the team should be allowed to dominate the discussion, no matter how insightful the contribution might be. Contrary views should be encouraged. At the same time, no member of the team should be allowed to "hide in the weeds". All team members need to respect each other and the contributions they are making now as well as potentially into the future.

Listening is an important ingredient in the team process. The team leader must be an excellent listener and must also demand that all team members listen to each other.

From time to time, a team buster shows up as part of the team. The team buster's motivation is often not known and not expected. Such a person must not be allowed to degrade the operation of the team. The team leader needs to try to get the team buster to change behavior. If this effort turns out to be futile, the team leader must have the power to remove the team buster from the team. This may be seriously traumatic. But experience shows that it is necessary. Decisive action is required in dealing with a team buster.

There is rich literature regarding the building of teams and what it takes to be able to serve as a team leader. Here are just two references that might help the reader to understand how to proceed with team leadership and productive behavior [1,2]. Some thoughts from an earlier work of this author [2] are included.

Project Management and Leadership

This author's perspective regarding project management and leadership starts with a short list of skills needed to be successful in this domain [1,2]. An excellent text on project management has been provided by Harold Kerzner over the years. Possibly the reader has a more interesting treatise on this subject, although Kerzner's book is comprehensive. This author's book on reengineering has an overview chapter on leadership that provides many perspectives on leadership from many leading researchers and observers of the scene.

References and Recommended Reading

1. Kerzner, H., "Project Management – A Systems Approach to Planning, Scheduling and Controlling", Third Edition, Van Nostrand Reinhold, 1989.
2. Eisner, H., "Reengineering Yourself and Your Company", Artech House, 2000.

30. LISTEN TO YOUR ELDERS

Let us start with what some of the elders had to say.

1. **A. D. Hall** [1]. One of the earliest purveyors of information regarding systems engineering was A. D. Hall [1]. He set forth some of the first principles when he was with Bell Labs, so it may be inferred that these labs were the spawning ground for systems engineering (SE). Topics of special interest explored by Hall include:
   a. A definition of systems engineering
   b. When and how SE is used
   c. The five phases of SE
   d. The theory of value represented by SE

2. **Simon Ramo** [2]. Well known as part of the TRW (Thompson-Ramo-Wooldridge) company. He was principal player in bringing systems engineering to the West Coast area and expanding with time the discipline itself. He was a leader in establishing systems engineering practices and methods. He served on various panels and received many awards for his contributions to the field. He expanded the utility and areas of application of the field at large. He was clearly one of the "larger than life" players that moved the field forward, both technically and from a business point of view.

3. **Eberhardt Rechtin**. We've visited this station before, and for a good reason. A conspicuous elder, Rechtin made seminal contributions to thinking about systems architecting [3] and doing the same with the organization [4]. He collaborated with M. Maier [5], expanding the knowledge base in systems architecting. He also held important positions at the Jet Propulsion Lab, the Aerospace Corporation (president), the USC, and Defense Advanced Research Projects Agency (DARPA). One can see from this short list and his writings that he played important roles in government, industry, and academia.

4. **Andrew Sage**. Andy Sage covered the waterfront in both systems and software engineering [6,7]. He served as dean at George Mason University, where he supported scholarship through coursework on systems engineering as well as what systems engineering is all about. He served as editor in chief of the *Systems Engineering* journal, an important springboard for research in the field. When the

annals of systems engineering are written, Andy Sage will be at the top of the list of significant contributors and supporters.

5. **Barry Boehm**. Barry Boehm, a leading software engineer, set the stage for a deeper understanding of SW economics through his constructions of Constructive Cost Models COCOMO I [8] and COCOMO II [9]. His real-world papers also helped deepen our comprehension of how SW engineering really works. Dr. Boehm also serves as a key researcher at the Systems Engineering Research Center (SERC).

6. **Yacov Haimes**. Dr. Haimes is an expert in risk analysis with major contributions in academia as well as classical risk assessment [10]. His book on risk modeling and management is a quite significant work, used by many when exploring matters of risk. He was a science fellow in the office of the president at the Software Engineering Institute at Carnegie Mellon University, and a professor of engineering at the University of Virginia. He is a fellow of seven technical societies and is a "national treasure" in the field of risk analysis and mitigation.

7. **Fred Brooks, Jr**. Fred Brooks is also a software engineer [11] and is credited with bringing the IBM 360 series to life. Fred Brooks is well known for his "mythical man-month" and his assertion that adding software engineers to a late project is likely to make the project even later. He followed this book with a second set of software engineering essays that provide even more insights into this field from a leading practitioner.

8–12. **David Walden, Garry Roedler, Kevin Forsberg, R. Douglas Hamelin, and Thomas Shortell**. These five authors are credited with having written the *INCOSE Handbook*, 4th Edition. This was an important milestone that highlighted the elements of systems engineering using the framework of Mil Std 15288. The latter emphasized processes, and it appears that these descriptions will carry the day for quite some time to come.

13, 14. **Blanchard and Fabrycky**. Professors Ben Blanchard and Wolter Fabrycky have represented strong foundations of systems engineering and analysis at the University of Virginia for many years. Their book is also a classic covering systems engineering and analysis as a field with completeness and clarity. They deal appropriately with the various forms of system design at its various layers. They cover economic evaluations at exactly the right level of detail. They look at design through the various lenses of the "-ilities", a unique approach that pays special dividends. This long-term collaboration of two giants in the field exemplifies the best in academic teaching. Both professors are now emeriti at the University of Virginia.

15. **Sarah Sheard**. Dr. Sheard is a leading researcher and consultant in the field of complexity. She did seminal work at the Stevens Institute of Technology and has been a major contributor to the field at the Software Engineering Institute at the Carnegie Mellon University. Sarah is well known for her exploration of principles for mitigating complexity in aircraft systems.

References and Recommended Reading

1. Hall, A. D., "A Methodology for Systems Engineering", D. Van Nostrand, 1992.
2. Ramo, S., "The Business of Science", Hill and Wang, 1988.
3. Rechtin, E., "Systems Architecting", Prentice-Hall, 1991.
4. Rechtin, "Systems Architecting of Organizations – Why Eagles Can't Fly", CRC Press, 2000.
5. Rechtin, E., and M. Maier, "The Art of Systems Architecting", CRC Press, 2009.
6. Sage, A., "Systems Engineering", John Wiley, 1992.
7. Sage, A., and J. Palmer, "Software Systems Engineering", John Wiley, 1990.
8. Boehm, B., "Software Engineering Economics", Prentice-Hall, 1981.
9. Boehm, B. "Software Cost Estimation with COCOMO II", Prentice-Hall, 2000.
10. Haimes, J., "Risk Modeling, Assessment and Management", John Wiley, 2009.
11. Brooks, Fred., Jr., "The Mythical Man-Month, Essays on Software Engineering", Addison-Wesley, 1975/1995.
12–15. D. Walden, G. Roedler, K. Forsberg, R. D. Hamelin, and T. Shortell, "Systems Engineering Handbook", John Wiley, Fourth Edition, 2015.
16–17. B. Blanchard and W. Fabrycky, "Systems Engineering and Analysis", Fifth Edition, Prentice-Hall, 2011.

31. LEADERSHIP

My time as a systems engineer was, partly, my time as an executive. This was a good time to observe my bosses and their leadership qualities. We had purchased a company by the name of Intercon Systems. I was soon appointed as president of that enterprise. I said to myself – well, here we are. It's time to really lead this company and "implement the theory".

So I went back to some notes and some writings [1] and tried to get a better handle on what it takes to be a leader in today's high-tech world. Here are some thoughts along these lines, trying to look straight ahead at what might be called the attributes or characteristics of today's leader.

Practical Visionary

The leader of today's high-tech company must be able to look ahead and see a definitive and very positive scenario for the company, based upon the practical actions that can and should be taken in the next three to five years. The visionary part is not a pipe dream. It's highly visible and within the company's grasp. But this is just one aspect of what a leader is all about.

Inclusive Communicator

The messages that are mastered by today's leader must go directly to the executive team as well as the rank and file in the company. There are no intermediaries, although the leader gets a lot of help from the next level of management. No one is left out of the communications chain. That's a ground rule that needs to be meticulously followed. And no backtracking from promises made.

Positive Doer

Many leaders go into a high action mode when they find themselves running a company. I actually slowed down, became more deliberate, and gave more thought to be positive as I met the people in the company, from secretary to vice president.

Renewing Facilitator

A big part of the leadership role in a company is to help others get their jobs done. This also means to help re-define what their jobs have become, as well as to breathe new life into the form and function of their daily activities.

Principled Integrator

Above all, the president sets the stage for success in terms of impeccable principles and modes of behavior. This is followed, just a step behind, by integrating the parts of the company that need to be in direct communication as well as action.

And to add to the mix, we try to remember what Peter Drucker said about leaders:

- "Managers do things right, and leaders do the right thing" [1, p. 179].

Reference and Recommended Reading

1. Eisner, H., "Reengineering Yourself and Your Company", Artech House, 2000.

32. NEW BOSS

The systems engineer tends to have a "new boss" each time he or she starts working on a new system. These relationships usually have a significant effect on one's career. You need to perform well for your new boss, even though there's a new set of behavior that one needs to grapple with.

As you experience new bosses in different project settings, you become "wiser" and more determined to do a better and better job. This is usually quite challenging, and you may need to make more and more determined efforts and adjustments. These adjustments will tend to correlate with who the new bosses are and their styles of management. Here are a dozen challenges presented to you by a new boss, and your awareness should be high as you see something new that requires adjustment on your part.

a. Your new boss – the micromanager
b. Your new boss – the heavy-duty authoritarian
c. Your new boss – the big-time planner
d. Your new boss – deeply organized
e. Your new boss – younger and less experienced than you
f. Your new boss – highly impatient for results
g. Your new boss – preferring action to just talk
h. Your new boss – the student of Myers–Briggs
i. Your new boss – the "close to the chest" person
j. Your new boss – the "all conversation is a progress report" person
k. Your new boss – the perfect controller
l. Your new boss – the consummate salesman

Your micromanager new boss. Stay out of his or her way to minimize them breathing down your neck

Your authoritarian new boss. Same as above to avoid getting a new assignment or a change of direction

Your big-time planning new boss. Draw up an early detailed new plan for your work that fits the overall project plan

Your highly organized new boss. Outorganize before he or she asks for your plan

Your young and inexperienced new boss. It's not about age; it's about the right ideas and the right well-considered actions

Your impatient boss. Wants results yesterday, so slow-roll as a counterforce

Your action-oriented new boss. Keep busy all the time, and never be found with your feet up on your desk, just thinking

Your boss who deeply understands Myers–Briggs. Take the test and discuss your results in relation to his or her results

Your boss who keeps it all "close to his or her chest". Ask a lot of nonintrusive questions

Your boss who listens carefully for progress. Give short progress reports

Your boss who likes to control. Consider giving up some amount of control to him or her

Your boss who likes to "sell" his position

Notwithstanding your boss's idiosyncrasies and tendencies, you might well consider how to interact with your new boss, from this third list:

1. Treat your new boss with respect
2. Listen carefully to what he or she has to say
3. Ask questions that tend to assist in problem-solving
4. Be careful not to threaten your boss's prerogatives
5. Stay one or two steps ahead in thinking, if possible
6. Learn the ins and outs of corporate interactions
7. Make sure (if possible) to make a friend out of your boss's boss
8. Do not out-play your boss, especially in a group setting
9. Volunteer to run meetings, but in a quiet, reserved way
10. Understand your boss's strengths and weaknesses, and act accordingly

Reference and Recommended Reading

1. Myers, I. B., "Gifts Differing", Consulting Psychologists Press, Inc., 1980.

33. TEAM BUSTERS

As a relatively young project manager I experienced an interaction that confounded and puzzled me. One member of my team seemed to attack and challenge me with respect to just about everything I said and did. He was quite disagreeable, and it looked like he was hell bent on deposing me as the project leader. After a while, I came to think of him as a team buster, trying to sabotage the team I was trying to build.

Nothing seemed to work as I tried, futilely, to bring him aboard as a productive team member. I was failing, and both he and I knew it. Here are some of the actions I tried, again without success:

a. Listen harder to whatever my team buster said
b. Give my team buster time to express himself
c. Acknowledge the usefulness of his criticism and ideas
d. Tried the opposite, from time to time, of items (a) through (c)

Nothing seemed to work as Mr. Team Buster was sniping at me just about all the time. After a while, I realized that I had a real "team buster" on my hands, an important lesson learned. But, what to do about it? I finally concluded that I didn't know how to "solve" this problem. Eventually, and in desperation, I finally moved this person off the project team. I basically "fired" the problem person, and without looking back. I felt right, and my intuition confirmed that I was on the right track.

So the answer seemed to be – get the team buster off the team. Nothing short of that seemed to work.

I also came into contact with a true "team buster" in two other situations. I had signed up for a Scott Peck-based seminar, exploring the ideas and behaviors of this well-known author of *The Road Less Traveled*. Karen was to be the seminar leader, and we were all supposed to meet at a designated time and space outside of Baltimore. As it turned out, Karen appeared, but more than a half hour late. About a dozen of us waited for her, mostly with a self-contained "wonder where Karen is" attitude. Finally, Karen appeared and apologized profusely for the very bad traffic tie-ups. The group reacted with approval and happiness that Karen was indeed in front of us and was ready to go. As we proceeded, however, one member of the group started sniping at Karen and berating her for being late.

"Can't understand how you could keep all of us waiting for so long", he said.

"We should be a lot further along, except for your lateness".

"Did you really study with Scott Peck?"

"These are not very interesting exercises".

Finally, I could not take any more of it. As a senior member of the guests, I spoke up.

Listen, Tom, I find your sniping at Karen completely obnoxious. She told us why she was late, and the traffic problem was not of her making. And she apologized, showing respect for the group. But your sniping is weird and off the page. So I ask you to stop doing it, and show respect for both Karen and the rest of the group.

He replied.

Whoever you are, you're completely out of line. I have every right to say what I'm saying, and every right to express my disapproval of Karen and what she's doing. I know how to present a seminar and she's not doing a good job of it. So please stay out of it, and stop showing disrespect to me and the other guests.

And then I replied with something like:

Looks like we're not going to get along during this seminar. For some reason, you don't want to accept Karen's lateness or her apology for being late. You've been attacking her for the past half hour, and I'm sure there's a hidden reason for the nasty and inappropriate behavior. But I'm going to call you on it, every time. So get yourself prepared.

"I accept the challenge", he said, "and look forward to having another enemy in my life".

And then, from another guest:

"Go at it", he said. "But I don't get any of it. All I want to do are the exercises that Karen sets up".

And so it went, for the rest of two days.

I suspect that Tom was a team buster who wanted to put Karen down and take over the group in some way or the other. That's what team busters do.

So if and when I run into potential team busters, I know the answer as to the lesson learned, at least for me. Out the door, never to return.

34. MEETINGS

Over the years, like many engineers of various types, I've attended hundreds (possibly several thousand) of meetings. For quite a few of them I've been the "boss". For others, I've been an observer or participant. You know, the usual.

If I've simply been a participant, then life would have been a lot easier. All I've had to do is pay attention and be there, at the moment. And when the time is right, answer a question or two that's been put to me. Sometimes the answer is "I don't know", or something like, "I'll try to get that answer to you by tomorrow, COB – close of business". It's often the best approach to delay or put off an answer when you really don't have one.

When my role at meetings has been the "boss", I've tried to make my attention span even more intense. Other folks at the meeting want to be listened to, and even heard. And beyond that, responded to. Preparation as the boss for a meeting can be extensive, dealing with such areas as:

a. List of agenda items
b. Notes on each agenda item
c. The current or new status of agenda items
d. Special dates and results
e. Roundtable query from all participants

It gets even crazier when there's some element of negotiation at the meeting, and you're the boss.

There are times when I've experienced what is called groupthink at a meeting. This is a particular form of dysfunctionality in which people do not wish to speak their minds due to the possible concern for what others will think of them and their answers. A more specific "definition" of groupthink can be taken to be [1]:

• A group process whereby people do not speak up for fear that they will appear to be out of step with the majority, or just look foolish for one reason or another.

Many books and articles have been written about meetings and how to deal with them. My lesson learned in this regard is to prepare well and try not to just "wing it". You may not be the smartest person in the room on any given subject, but with the appropriate level of preparation, you may appear to be extremely well-informed on the topic at hand. Usually, that's more than good enough. Going beyond that, one may try one or two methods that are reported in the literature. A particular viewpoint on how to improve meetings is provided as "The Interaction Method" [2]. This method includes four participants in any and all meetings:

1. The "boss" and chair of the meeting who moves activities and problem-solving activities along,
2. The *facilitator*, who helps each group member express and clarify ideas and approaches but does not "take over" the discussion,
3. A *recorder*, who keeps accurate and unbiased records of what has transpired, and
4. The *individual member* of the group/team. It is claimed that increases in productivity of up to 15% have been achieved using this method.

This number is large enough to suggest that the method be used in the systems engineering community, at least on an experimental basis.

A Systems Engineering Meeting

If the meeting is in the context of a systems engineering project, then take a look at a systems engineering book that highlights the topic "systems engineering management". Topics that come to mind in this connection include:

a. Special technical requirements
b. Special contractual requirements
c. System functional decomposition
d. System architecture
e. Integration – problem areas
f. Test results and plans
g. Modeling and simulation results
h. New inputs from customers or special stakeholders
i. Percent completion – cost, schedule, performance

As a final reference let us take a quick look at some "meeting" suggestions from Simon Ramo [3], a super-engineer as well as a super-businessman. He claims to have attended some 40,000 meetings of various types:

a. Abolish unnecessary meetings.
b. If there is a Machiavelli-type in the group, be careful about what transpires (Machiavelli has his own agenda).
c. The chair takes the role of the leader, and also as "chameleon".
d. Be thoughtful and careful about seating arrangements (there is such a thing as bad seating).
e. Look out for the "Must-Win" debater.
f. Watch for and try to specify attire.
g. The worst personality type that he encountered is the MDRSSA, the "Multi-Dimensional Really Smart Smart-Ass".

References and Recommended Reading

1. Sage, A., and W. Rouse, "Handbook of Systems Engineering and Management", John Wiley, 1999.

2. Doyle, M., and D. Straus, "How to Make Meetings Work", Jove Books, 1976.
3. Ramo, Simon, "Meetings, Meetings and More Meetings", Bonus Books, 2005.

35. MYERS–BRIGGS

Looking back some 50 years in the SE world, I found two instances in which I ran directly into the personality measurement part of the world, sometimes called psychometrics. Both were similar in that they provided some type of lessons learned. Both had some food for thought in an unexpected way.

In one case, I was teaching a course in Project Management (PM) as part of a GWU program. The Myers–Briggs Type Indicator (MBTI) was part of the curriculum, and so I took the "test" along with some 25 students. I found out that I was an INTJ, which surprised me a bit. First of all, this profile may be contrasted with its polarities in that

I (Introvert) is opposite to E (Extrovert),
N (Intuitive) is opposite to S (Sensing),
T (Thinking) is opposite to F (Feeling), and
J (Judging) is opposite to P (Perceiving) [1].

This structure is what is known as the Jung typology.

This INTJ indicator apparently is present in only about 2.1% of the population, which is relatively rare (see Table 35.1).

"Strange", I thought, "but likely to be a real challenge". And so it has been. See if you can find any "strangeness" relative to your MBTI from the rest of the types and frequencies, as below:

TYPE	PERCENT (%)
ISFJ	13.8
ESFJ	12.3
ISTJ	11.6
ISFP	8.8
ESTJ	8.7
ESFP	8.5
ENFP	8.1
ISTP	5.4
INFP	4.4

So I discovered that "the NTs tend to be logical and ingenious and are most successful in solving problems in a field of special interest" [1]. For introverted

TABLE 35.1   Lowest Seven
MBTI Values (%)

ESTP	4.3
INTP	3.3
ENTP	3.25
ENFJ	2.5
INTJ	2.1
ENTJ	1.8
INFJ	1.5

thinking, the "goal is to formulate questions and create theories" [1]. For INTJ and INFJ types, some citations include [1,112]

- Finding new pathways
- Being motivated by inspiration
- Looking for deeper meanings
- Extreme discontent with a routine job, and its implications

INTJs are taken to be the most independent of the 16 basic types, and this type of behavior will show itself when an INTJ is part of a team.

In another adventure with project management, I gave and took a "test" that yielded the following four values:

Action (*)	13 (my score on a scale of 10-10-10-10)
People	11
Process	6
Idea	10

(*) *action* is to very much enjoy getting it done; *people* is to look for positive interactions with people; *process* is to hold on to ways for the person and the team to behave, once established; and *idea* is to formulate new ways of behaving based upon the new and clever idea.

The part of this that surprised me was my relatively high "action". I did not think I was action-oriented but was pleased to find out that this might not be true. Another lesson learned and not forgotten.

Reference and Recommended Reading

1. Myers, I. B., "Gifts Differing", Consulting Psychologists Press, Inc., 1980.

36. BECOMING A HI-TECH MANAGER

For the systems engineering path that so many of us travel, there comes a time to decide whether or not to become a manager. It's a difficult transition for many. I decided that the answer was "yes", although the project size was so small that I barely noticed my new role.

There were two cases I can note. One was as project manager of a program for NASA on the Nimbus meteorological satellite. The other was for the Federal Aviation Administration (FAA) on a radar quality control experiment. The Nimbus project was about five people, and the FAA was just me and the other guy. Starting out small is an exceedingly good idea. You're nimble, and there's not a lot of yelling and screaming.

So what's the lesson learned in this endeavor. As you might expect – it's simply the elements of doing the project manager job. Sounds simple, but has many dimensions that you might not have anticipated. Examples?

1. Progress reports
2. Briefings
3. Review of all reports
4. First-line interaction with customer
5. Lead on possibly all technical matters
6. Run all meetings

There are others, I'm sure.

Now that it's close to 50 years later (with respect to my first project manager experience), it's time to look at what's in the literature on this matter. In this regard, let me start with what I had put in the literature a few years back. This singular source [1] looks briefly at the skills required and the specific steps that are suggested.

Skills Required

In a short exposition, this author identified five essential skills for a high-tech manager [1]. These are briefly cited and discussed below.

1. **Problem solver.** Above all, this person needs to be able to address problems and find real, practical solutions. This includes individual as well as group problem-solving.
2. **Contingency planner.** This item deals with the fact that this person often needs plan A, Plan B, and Plan C to be successful over the long haul.

3. **People-oriented communicator**. Two ideas are represented here. The first is that today's high-tech management must be comfortable dealing with people on a continuing basis. The second is that his or her communication skills must be naturally superior.
4. **Team builder**. A lesson learned and articulated early in this treatise has to do with building a strong team. This is more than a random collection of engineers who show up at systems engineering meetings of one sort or another. This type of team exhibits high-performance and high energy, and they work together in an especially productive manner.
5. **Technically competent decision maker**. Here again, this implies two features: technical competence explains itself and decision maker confirms an ability to clearly resolve both technical and management problems.

Specific Steps

Moving from a team member to a high-tech manager requires a series of steps. These are briefly cited below:

1. **Make a conscious decision and commitment**. You basically need to say "hello" to your new job as a manager.
2. **Obtain additional training and education**. This includes formal courses leading to bachelor's and master's degrees as well as special certificate programs.
3. **Practice managerial skills**. Look for occasions in which you can engage in specific problem-solving sessions. Volunteer to run project review activities.
4. **Study and talk to managers in your company**. Target and engage with specific people.
5. **Seek and accept a manager position**. Sometimes it happens quietly and with no effort. Other times it requires a more proactive set of actions.

Reference and Recommended Reading

1. Eisner, H., "Reengineering Yourself and Your Company", Artech House, 2000.

37. DEALING WITH YOUR CUSTOMER

First, let's set up a reasonable scenario.

You're a middle manager, and a couple of years ago you wrote a spectacular proposal and won a government contract with your current customer. The contract is significant in size, and you're the designated project manager both in name and in deed. The re-compete is due in a couple of years, and it's up to you to assure a win in that competition. If you lose that re-compete, you may be out of a job, but if not that, it's a couple of rungs down on the ladder from which you may never recover. So you're strongly motivated to be successful for your individual career and your placement within the company.

You maintain day-to-day contact with your customer. He depends on you to respond successfully to issue task orders on a timely basis. He is able to send you fixed price orders that do not require you to compete. Some orders, however, are competitive with other contractors. So, even though you won this contract, you still need to be able to compete, from time to time, on individual task orders.

You approach this challenge by trying to create a trusting relationship with your customer, and always doing A plus work. Building trust takes time, and you need to demonstrate that you can be trusted with important assignments. You also look for ways to build trust between you and your customer. It all takes time, the right behavior and a lot of patience.

Going to Lunch with Your Customer

Your customer invites you to lunch for a "working" session to review the status of various task orders. When you're finished, your customer makes clear that paying for his lunch would be ok, do you do this? The answer is a definite "no", since such action is likely to be illegal under the terms of the contract. You establish this boundary from the beginning. No free lunch and no free anything. How to do this? One way is simply to start out with "separate checks please". This is not easy to misunderstand.

Issuance of a New Task Order

Your customer asks you (your company) to respond to a new task order. You decide that this is a good time, since it is non-competitive, to manage expectations. You follow up with a "try to under promise and over deliver" approach. The customer is clearly very pleased to get the results early and of high quality. This all helps to build trust and a stronger relationship.

Quick Response Capability

On a Friday at noon, your customer calls with an urgent request. He needs a "Powerpoint" presentation of an important topic by noon on Monday. Are you willing and able to respond? It clearly means weekend overtime work for you and a couple of members of your project team. You decide "yes", and so you and your project team are verifying that you are willing and able to do these possible weekend fire drills. This is a service beyond the call of duty and is usually much appreciated. You are beating your competitors, hands down.

A Truthful Interchange

During one of your lunches, your customer lays out his approach to a particular problem he is having. You think that his approach is basically flawed, and therefore you cannot respond positively. Your customer is looking for support, but you find that you're unable to take such a position. You think long and hard about this situation and decide that all you need to do is be honest. You will do that by suggesting an alternative approach.

"Have you considered …?" you say, and give your customer a chance to recover and move on with his thinking.

"Thank you for your thoughts on this matter", he says. And you have gained some points by an honest response that moved the problem-solving forward, and with due consideration.

The Re-Competition

You approach your re-compete with confidence, but not over-confidence. You take nothing for granted as you write a comprehensive well-thought-out proposal. You cover the bases and provide alternatives to allow your customer to let more than one contract, if desired. You recognize that your customer needs to play everything by the book which you have facilitated by suggesting alternatives. You ultimately win the re-compete, which is a win-win for all parties. Your boss comes in to congratulate you on a job well done. This contract is now a "cash cow" for the company, and you're in charge. You've done a great job by establishing a new LOB (line of business), and you've done it the right way. You've built a solid professional relationship of trust with your customer. The relationship is characterized by:

a. Mutual trust,
b. Well-established boundaries,
c. Solid technical expertise,
d. Responsiveness above and beyond,
e. Following the procurement rules, and
f. Mutuality of interest.

38. INTEGRATION

Integration is a procedure by which the systems engineer (or team) brings together the various parts of the system. We seek a special way to do this that is efficient and cost-effective. If the subsystems are designed to be readily integrable (is this a word?), then the integration task tends to be facilitated. If they are not so designed, the integration can be difficult to impossible. Are there some ground rules for proceeding with this set of tasks? Let's put a few of them on paper and hope that they'll be helpful. Several items on this list can also be found in a textbook [1] by the author.

1. *Always architect at least two systems from which to choose a preferred system; the recommended number of alternatives is three*
2. *Do not conceive of and then integrate all stovepipes; integrate what it is that is provably cost-effective to do so*
3. *Assure that all members of the system integration team have specific experience with a significant systems integration activity*
4. *Consider technology insertion as a definitive task pertaining to system architecting*
5. *Accept evolutionary design and "chunking" (of software) as part of the integration process*
6. *Use reuse sparingly and as a part of your plan to build compatible software*
7. *Confirm that the project budget and timelines are sufficient to execute the complete integration activity*
8. *Attempt to reduce complexity in very specific ways, especially with respect to software*
9. *Treat requirements as tentative and in need of formal confirmation*
10. *Follow all acquisition system ground rules that pertain to this type of system*

Reference and Recommended Reading

1. Eisner, H., "Essentials of Project and Systems Engineering Management", Third Edition, John Wiley, 2008.

39. HALL, GOODE, AND MACHOL

Now let us take a quick look at how systems engineering was described in the very early days. From that we might gain some insight into why we no longer use these descriptors and have moved on to better ones in sizing and shaping the nature of systems engineering.

In particular, our quick look will involve A. D. Hall [1], Goode, and Machol [2].

A. D. Hall [1]

Hall's treatise is given the name "a methodology". It is not a coherent method, but it is a very interesting set of very relevant topics. It's a tour de force, including many subjects not found in today's systems engineering books. His top-level structure of systems engineering leans heavily upon planning, in particular program planning and project planning. The latter itself is composed of exploratory planning and development planning. Moving from this abundance of planning, we come to the "action" part of Hall's structure, namely, studies during development and current engineering. Certainly, in this presentation, planning is too extensive, and engineering too depressive.

However, very interesting chapters emphasize such topics as:

a. Decision making and games
b. Functional design
c. Synthesis
d. The theory of value
e. The role of measurement

In essentially all of Hall's books, we see little of requirements (but we do see "needs"), life cycle, and process. Hall had his own vision, and this author commends him for his cogent articulation of that broad and multi-faceted vision.

For those of us in the systems engineering community, it would do well to consider how to include some of that vision in our future renderings of systems engineering.

Goode and Machol [2]

Goode and Machol's classic text on systems engineering leaves us much to think about, even though it is quite different from today's notions. It starts with an emphasis on complexity and moves quickly into probability theory and applications. After a stop with computer concepts, it considers the basics of game theory, linear programming, group dynamics, and cybernetics. It finishes with information theory, servomechanism theory, and human engineering. Not exactly today's primer with which to gain certification in systems engineering. But not to be entirely neglected by today's systems engineers. And still, no exposition of "processes".

Machol's View [3]

Back in 1965 Robert Machol undertook a project to document his conception, along with others, of the field of systems engineering. The introductory chapter is entitled "Methodology of Systems Engineering", with the following component parts:

- Definition
- Anatomy of systems engineering
- Principles of system design
- The systems viewpoint
- Operations research
- The complete systems engineer

We note the emphasis on systems design, overall systems, and operations research. This perspective, of this author, is entirely appropriate all these many years later.

Dislocations occur, however, almost immediately, as Machol and his 58 contributors move into such topics as the ocean, land masses, the atmosphere, and astronomy. Then Dr. Machol brings in some subjects that are often seen as electrical engineering topics, such as analog circuits, transformational calculus, communications engineering, radar, and satellites. A movement in the direction of computer science brings computer system design and digital circuits and logic into play.

A major topic given the name "system theory" has logical elements such as information theory and game theory. Simulation is included and carries forth up to today.

This monumental work from Machol and colleagues makes its own contribution and is both broad and deep in its approach. It is interesting to look at systems engineering as seen by Machol in the 1960s and what it has evolved to in the form, for example, of Mil Std 15288. Another point of departure is the *INCOSE Handbook*, Version 4. Could you have predicted this evolution, or even a small part of it? Your thoughts?

References and Recommended Reading

1. Hall, A., "A Methodology for Systems Engineering", D. Van Nostrand, 1962.
2. Goode, H. and Machol, R., "System Engineering", McGraw-Hill, 1967.
3. Machol, R., W. Tanner, Jr., and S. Alexander, "Systems Engineering Handbook", McGraw Hill, 1965.

40. MAN VS. MACHINE

Some years ago, this author was engaged as a contractor to support the development of a three-axis stabilized meteorological satellite system. One morning, the deputy program manager came to my office with a concern of his. He was thinking about the paths of the satellite and the boost vehicle and the possibility that they might collide after some number of orbits of each. Could this happen, and are you in a position to analyze this type of situation?

I accepted the challenge and went back to my home base. After a couple of hours thinking about the problem, along with my book on orbital mechanics, I decided to call our chief scientist, Dr. K, to see what he thought about the overall problem. He invited me to his office, where I went just about immediately.

"Yes", he said, "I think I can work this problem, and give you an answer in a couple of days".

Dr. K was a man of his word and had never failed me in the past.

The next day I wrote up the easy part of the problem and solution and awaited word from Dr. K. Sure enough, by noon of the following day, I got his call and made it up one floor to his office. He showed me his work, emphasizing a polar plot of the paths of the satellite and the boost vehicle, as a function of orbital number. It showed the miss distances explicitly, and one could see no evidence of a crash between the satellite and the boost vehicle.

I was impressed, one more time, by Dr. K's analytic skills and how fast he was able to bring them to bear upon a problem.

I called the deputy program manager with the news, and by the next day I showed him the results in graphical form, just as Dr. K had constructed them.

"Terrific work", he said, and then confessed that he had given the same problem to one of his other contractors, the one who was the official keeper of the computer-based satellite orbital model.

"I don't have their answer yet", he said, "but I was promised an answer in two weeks".

I noted the difference – our answer in 2–3 days and the computer-based model in 14 days or so. There it was, a concrete example, if you will, between man and machine. And for whatever reason, man was to be the winner.

A couple of weeks later, I received a call from the deputy program manager.

"Just got the results from our computer-based model, and they confirm your answer".

"Glad to hear it", I said. "A very good answer and resolution". He must have heard how happy I was with the entire episode.

Dr. K was worth his weight in gold. Never underestimate the pure power of the human mind. A lesson learned, for sure.

Miscellany 5

41. REDUNDANCY IS IMPORTANT AND MAY BE CRITICAL IN CERTAIN SYSTEMS

Redundancy is a means by which one increases reliability, usually at the expense of space and/or weight, or both. I had an experience some years ago that perhaps will illustrate the point.

I was on a team consulting with a program manager for the Nimbus meteorological satellite. Our job was to look at the Nimbus design and carry out what at that time was called a design review and reliability assessment. I was "responsible" for the power supply and also the overall team activity. I came to the conclusion that there was a real danger (risk) associated with the power supply in the form of failure of the drive motor for the solar panels. This drive motor was a new design and failed in space after a short time. This failure meant that the solar panels would not be oriented properly with respect to the sun, and so the spacecraft was soon a hunk of junk in space.

And this was what was called a "single point" failure.

This "incident" stuck in my mind, and in all future efforts of this nature, I always looked for the possibility of single point failures and considered how to fix them.

One way to do so, of course, is through redundancy. That means that two failures need to occur in order for the overall system to go down. Here's a quick look at the relatively simple mathematics of a parallel redundancy configuration so that we can see what happens, quantitatively.

Let there be two identical units in parallel, and with reliability R. R is the probability that each unit will not fail (will continue to work properly). Then, the failure probability for each unit is $(1 - R)$. The reliability of the redundant configuration is therefore the likelihood that at least one (one or more) will not fail. That equation can be shown to be:

$$P = 1 - (1 - R)(1 - R) = \text{reliability of parallel configuration}$$

This reduces to $P = 2R - R^2$

Looking at some numbers, we see below a range of P for different values of R. The parallel redundancy certainly helps and does so dramatically.

UNIT RELIABILITY (R)	PARALLEL REDUNDANT RELIABILITY (P)
0.5	0.75
0.6	0.84
0.75	0.9375
0.95	0.9975
0.98	0.9996

Even if we start with a unit reliability of as high as 0.98, the redundant configuration increases the overall reliability to 0.9996.

Moving to another case, the reader might recall that NASA also had a quite serious single point failure on a manned mission known as Challenger. That was the so-called o-ring problem. Apparently, a critical o-ring froze and broke, constituting a single point failure that destroyed the mission and killed everyone aboard – an enormous tragedy. The crew was seven people: five NASA astronauts, a payload specialist, and a school teacher from the civilian world.

The Rogers Commission was established by President Reagan to investigate this accident. There were many findings, but the key problem was the failure of the o-ring. Dr. Richard Feynman, a Nobel laureate physicist, was on this Commission. He demonstrated the problem, and wrote about it in one of his books [1]. He showed, in a unique as well as simple construction, how the o-ring was likely to have failed. This was a powerful and convincing argument, and one that was needed at the time. The report of the Rogers Commission is excellent reading, showing how there can be many points of view expressed by a variety of people trying to reconstruct the truth about an accident. It would seem to be an open and shut situation, but apparently it was not. This "simple" world of ours can be quite complicated at times. Perhaps that is why "risk" is so difficult to assess. We have a major issue facing us when we might wish to make some changes late in a program.

This brings me back to the Nimbus drive motor failure and a discussion I had with the program manager in a mission postmortem. We went over our report, which cited the possibility of a single point failure of the solar panel drive motor. He acknowledged that we had made an excellent point, but from his perspective:

a. There were several other "problems" that were identified, and it was not clear as to what the priorities for fixing needed to be
b. We were quite late in the program, and a change would have meant considerable (and unacceptable) time delays
c. His key senior people did not support any changes as they approached the launch date

Management has to make these kinds of calls, and they live with the consequences. Often, there are regrets. But there is usually quite good justification for action, or lack of it.

Reference and Recommended Reading

1. Feynman, R., "What Do You Care What Other People Think – Further Adventures of a Curious Character", W. W. Norton, 1988.

42. RECHTIN'S HEURISTICS ARE BRILLIANT AND NEED TO BE STUDIED AND FOLLOWED

We cannot do better, as systems engineers, than to follow the advice of Eberhardt Rechtin. He was a "master" engineer, with an illustrious career in industry, government, and academia. He played a major role at the Jet Propulsion Lab, the USC (as a professor), and also as president of the Aerospace Corporation. He also held the position of director of DARPA, the Defense Advanced Research Projects Agency.

Dr. Rechtin wrote a seminal book [1] dealing with the subject of architecting systems. This was the first in this field, which amplified his approach with a follow-up book with Maier [2].

A distinct feature of the first book on architecting was his Appendix on heuristics. This appendix set forth a list of "rules-of-thumb" that Dr. Rechtin constructed from his many years of building systems. This author sees this list as lessons learned that Rechtin is passing on to other systems engineers in the field. Here is a list of ten of these heuristics to be read and re-read by our community of systems engineers.

1. **KISS (keep it simple st..id)**. Dr. Rechtin emphasized simplicity in his designs, believing that this approach would bring dividends in terms of decreased cost and increased performance. How does one do that? Answer – follow Dr. Rechtin's suggestions.

2. **Keep system requirements under challenge**. Keep questioning your list of requirements that tend to drive the overall system design, and modify them when needed.

3. **Important software mistakes are made on the first day**. One may infer from this that we make errors early in our software architecture, and we need to pay attention to this activity.

4. **No system can be optimal for all parties**. We do our best in many domains but cannot succeed in all of them simultaneously.

5. **For new systems, expect the unexpected**. New problems arise in building large-scale systems, just about all the time.

6. **The design team cannot avoid re-design**. Look for design weaknesses, and take actions to avoid them as early as possible.

7. **Maintain options as long as possible**. Keep from finalizing system design so that changes can be made without too much pain.

8. **Try to assure minimum communication between subsystems**. Keep the inter-system message flow down from proliferating.

9. **Choose among alternative architectures**. This implies that you have constructed alternatives, which is good engineering practice (note AoA within the DoD).

10. **Recognize Pareto's law**. Keep this 80–20 law in mind and behave accordingly (if you've forgotten the law, it's that 80% of the significant work in an organization is generally done by 20% of the people).

References and Recommended Reading

1. Rechtin, E., "Systems Architecting", Prentice-Hall, 1991.
2. Rechtin, E., and M. Maier, "The Art of Systems Architecting", CRC Press, 2009.

43. MISTAKES

So there came a time when Xerox was the most successful enterprise in the country. The copier company was led by Joe Wilson, its president, and Sol Linowitz, its chairman. Their clever capture of key xerographic patents and strategic alliances with Rank (UK) and Fuji (Japan) put Xerox in a key position in this country's technology marketplace. That allowed them to set up Xerox PARC as a high-technology leading edge company. This enterprise soon took charge of SDS computer and made several poor decisions in this respect. SDS was an asset that was squandered, and the story behind that is not easy to figure out, at least in terms of corporate activities. This was not the only mistake made by Xerox; we cite here a failure to understand how it continued to benefit from Xerox PARC and instead invested in real estate, which turned out to be less than lucrative [1].

When Anne Mulcahy took over the reins of Xerox in 2001, it had a $273 million loss and a stock that had dropped some 92% in less than two years [2]. Stockholder value decreased by 38 billion. Its bonds were rated as "junk" by Moody's. One might say that the company was barely breathing. But Mulcahy forced major cuts in the company's cost structure and also in the year-by-year budget. Many of her moves brought the company back from "from the edge" to live another day.

Another quite serious mistake in the computer business was made by the ex-MIT engineer-owners of the DEC (digital equipment corporation). They had a relatively solid line of business with their PDP and VAX series and sold it to COMPAQ. Sales declined as did the value of DEC stock. Whose mistake was accounted for here? Not clear – several folks to "blame".

Wang Labs was highly successful as it essentially captured the word processing market, only to give it up to "open" systems word processors (like WordPerfect, WordStar, and others) and computers. It went bankrupt in August 1992, cutting some 5,000 jobs and never did recover from its early days of astonishing success.

Another computer-related story and mistake is connected to Apollo computer, a powerhouse workstation manufacturer that competed strongly for the preferred government workstation. They lost that competition to a team of UNISYS and SUN microsystems. Apollo's president said at the time, from his perspective, "it's ours to lose – and they did". SUN continued onward and upward – and Apollo more or less disappeared from the scene.

Then we get to IBM and their adventure with the IBM PC. They went to Microsoft to obtain an operating system for their PC and ultimately made a sweetheart deal with Microsoft. The software company wound up owning the operating system and used it to its advantage for many, many years. How did that happen? Some say it was due to the fact that IBM did not properly value software. We presume that by now IBM has figured it out – and learned "how to dance". It took quite a while, but they survived that and other misadventures and appear to be alive and well today.

A Footnote to the IBM – Microsoft Story

In and around 1979, Microsoft said "no", declaring that they were not in the business of constructing operating systems. However, Gates referred IBM to a friend in another company, a friend definitely in that business. His name was Gary Kildall, and the company was Digital Research. So IBM went off to visit with Kildall, and when they got there, they found that Kildall was off flying his plane, and he was represented by his wife for the meeting. She said that Digital Research was indeed in the "operating system" business, as, for example, they had built CP/M for PCs. When asked by IBM to move forward with a non-disclosure agreement, Kildall apparently claimed that she did not have the authority to execute such an agreement. So IBM simply left and went back to Microsoft. Having done a favor for a friend, Gates now felt that he could now say "yes" to a deal offered by IBM. That's exactly what he did, which led to Microsoft going off to purchase a system called QDOS for about $50,000 from a company called Seattle Computer Products. That was a starting point for Microsoft in their new adventure with IBM, and their monopoly-building with DOS, MSDOS, and Windows. Not bad, eh? A clear turning point for both IBM and Microsoft in the field of operating systems. And Microsoft had the powerhouse known as "big blue" behind them for each and every operating system they sold.

Mistakes in the software arena as well as segments of the computer world appear to be plentiful. But fortunes have come and gone, and many are still "cashing checks", and also still making mistakes. Are there some lessons to be learned from all of this? The answer is "yes", and we can cite one such lesson that stands out above the others. And that is – when you've made your first billion, pause for a while and consolidate. It's time to stop running free, and count your blessings as you play ball with the government. Even Facebook is still trying to figure it out.

References and Recommended Reading

1. Smith, D. and Alexander, R., "Fumbling the Future", toExcel, 1999.
2. Eisner, H., "Topics in Systems", Mercury Learning and Information, 2013.

44. COST ESTIMATING

An important lesson learned has to do with the often difficult field of estimation, particularly software cost estimation. Barry Boehm has given us some definitive guidance with COCOMO I [1] and COCOMO II [2], but both require inputs which, of course, are estimates.

So let us set up a scenario in which lead software engineers are sitting around a table trying to estimate the cost of a software project. Let us use COCOMO I to gain some insight into the nature of the problem we are facing. This team looks at the system specs, and they come up with an initial estimate of 100,000 delivered source instructions (DSI). With this value, we now proceed with several COCOMO I calculations as:

$$\text{Person-months}\left(\text{PM}\right) = 2.4\left(\text{KDSI}\right)^{1.05} = 2.4\left(100\right)^{1.05}$$

$$= 2.4\left(125.9\right) = 302.1 \text{ person-months}$$

From this estimate, we now compute the development time as

$$\text{TDEV} = 2.5\left(\text{PM}\right)^{0.38} = 21.9 \text{ months}$$

The productivity and full-time equivalent staff are then:

$$\text{PROD} = \text{PRODUCTIVITY} = \text{KDSI/TDEV}$$

$$= 100/21.9 = 456 \text{ KDSI/TDEV}$$

and

$$\text{FTES} = \text{PM/TDEV} = \text{FULL-TIME EQUIVALENT STAFF}$$

$$= 302.1/21.9 = 13.8$$

We note that only one estimate of 100,000 DSI gave us the next four estimates of PM, TDEV, PROD, and FTES. Now there's efficiency for one.

The software engineering team is asked to stare at these numbers for a while and be prepared, one by one, to comment. Most of the comments center upon the uncertainty in the overall process and the set of "output" estimates. What to do next? The team expresses what it considers to be a common perspective:

"Let us push for numbers that represent the overall opinion of the team".

So each and every person on the team was asked for an estimate. "So what is your estimate for the number of Delivered Source Instructions?" And, of course, we get a spread of numbers and know not what to do with them.

- What does the systems engineering team do when a single parameter is estimated by more than one person? And how does the team approach the overall problem of cost estimation? [3]

References and Recommended Reading

1. Boehm, B., "Software Engineering Economics", Prentice-Hall, 1981.
2. Boehm, B., "Software Cost Estimating with COCOMO II", Prentice-Hall, 2000.
3. Mislick, G., and D. Nussbaum, "Cost Estimation", John Wiley, 2015.

45. GENERALIZE

A story out there in the literature involves a strategic planning session of the American Association of Railroads (AAR), around the turn of the century (1900). A key question put to the attendees was:

We are the AAR and what business are we in?

The answer was:

We're in the railroading business, of course.

This answer was roundly accepted as the correct one. But we note that there was not one contrary view with an answer that said:

We're in the transportation business, of course.

This difference in the answer as well as perspective, it is told, led to the result that the Railroaders were not, by and large, leaders in the formulation of the aviation industry.

This "story" can be used to suggest that a generalization is a powerful tool in the world of strategic planning and thinking. In that world, generalization may lead to new systems, products, and services. It may lead to breakthroughs in terms of lines of business and positioning in the marketplace.

All it may take, as suggested, is a voice of generalization at the right time and in the right place.

I offer another example from my background that ultimately led to rather positive results. Our company was doing well in the business and high-technology areas of NASA's Goddard Space Flight Center (GSFC). We were certainly getting our share of contracts when we sat down one year and took strategic planning a bit more seriously. There it was:

• What business are we in?

One answer (not the only one) was "we're in the space business". This answer ultimately led us to another (and different) customer, namely, the Air Force. We ultimately won contracts with this important client in quite distinctive areas. These led to specific efforts with the Air Forces Consolidated Space Operations Center program and the Strategic Defense Initiative (SDI) as well as missile defense programs. Thinking in broader terms eventually paid off. Generalizing turned out to be a good thing. Problem-solving for one "space" customer led to contractual work of substance and high quality for another "space" customer. It was a success story that happened but did not happen overnight or without considerable effort.

I put this on my list of lessons learned.

46. RISK ANALYSIS AND MITIGATION

For this author, this statement is real and needs to be instantiated in action. And what is the "it"? Let's use the name "risk assessment and mitigation" (RAM). And let's remember to do the last part with as much focus and energy as the first part, namely, mitigation.

RAM can be viewed from several perspectives. A top-level view is to suggest that it can have four elements:

a. Performance risk,
b. Cost risk,
c. Schedule risk, and
d. Societal risk.

These are often quantified, representing the likelihoods of (1) meeting performance requirements, (2) staying within budget, (3) satisfying time milestones, and (4) having little to no negative effect on society.

As with many fields, there are many "gurus" in this one. Two of them are:

a. Yacov Haimes [1]
b. James Reason [2]
   The latter researcher is credited with having constructed the so-called Swiss cheese model, which deals with layers of defense that are set up as the essential part of risk mitigation. These layers can be thought of as independent ways to block a bad intrusion into the system in question. The "Swiss cheese" model has many areas of application, including systems engineering, cyber-protection, healthcare, and warfare (to name just a few.) By way of illustrating the latter area, imagine the old days of battlefield or trench warfare. The generals established these layers, each one a defense against enemy penetration. If the penetration gets through the first layer, it goes on to the second, and so on. One can readily see that multiple layers will reduce the risk of ultimate penetration.

Viewed from a probability perspective, let us assume that

P = the probability of layer failure (fails to stop the penetration), and
Q = the probability of layer success (stops the penetration). From this simple assignment we see that

If the failures to stop the penetration are all the same, namely P, then the probability of failing at each of N stages is simply

$$P \times P \times P \dots P = \left(P\right)^{N}$$

If P, for example, is (0.1), then the overall failure to penetrate four successive layers is $(0.1)^4 = 0.0001$

We can change the "resolution" of this model such that P is 0.01 or any arbitrary value, but the overall concept remains the same. It all depends upon how thin you like your Swiss cheese sliced, if you will.

Take Your Pick of Serial and Parallel Configurations

Equipment 1	Equipment 2
Reliability 1 = R (1)	Reliability 2 = R (2)

Serial Reliability:

R = Overall Reliability = $R(1) \, R(2)$

Parallel Reliability: (simple redundancy)

Diagram:

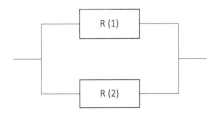

Overall Reliability = 1 - not R (1) not R (2)

When reliability is exponential, $R = e^{-ut}$; where u = failure rate and t = time.

References and Recommended Reading

1. Haimes, Y., "Risk Modeling, Assessment and Management", John Wiley, 2009.
2. Reason, James, "Human Error Models and Management", British Medical Journal, 320(7237), 768–770.

47. CHANGE, OPTIONS OPEN, AND ITERATION

There are several "change" concepts and contexts in systems engineering. Each has to be considered and reckoned with in order to do the best possible job on a real project or program.

One context is connected to the experience I had discussing a configuration change, after the fact. This is not an unusual context and is played over many times on a typical project. In this case, we were discussing a failure on the solar panel drive motor of the Nimbus satellite project. The drive motor had failed, which soon meant that we lost power due to the poor orientation of the solar panels. The drive motor had been tested, but possibly not enough.

The drive motor was not in a redundant configuration. The drive motor was clearly a single point of failure risk. We were, in principle, trying to eliminate all single point mission failure situations – big red flag with the solar panel drive motor.

I can't recall the details of that discussion, but we very likely did not feel we were in a position to change the configuration with the solar panel drive motor and therefore did not do so. In principle, we could have made such a critical change, but again, we did not do so. So the question arises:

- When do you make a system design change, and when is it out of bounds to consider a change?

Another much more well-known case is that of the "o-ring" problem. I'm not in a position to re-hash all the arguments for this complex case, but an a priori change was on the table and it did not occur. So the question is:

> What made change possible, and what, in effect, made change impossible at that time?

If we generalize a bit we come to the notion of building more redundancy or design change into a configuration. This increases reliability but comes with a price, often a severe one, in terms of space and weight. When is it time to consider such a change having to do with various kinds of single points of failure?

Rechtin's Options Open

Related to this issue is what Rechtin has suggested in his "keeping options open" heuristic [1]. Here, Rechtin sets forth the following as a good practice heuristic:

- Build in and maintain options as long as possible in the design and implementation of complex systems; you will need them.

As usual, Rechtin has good words for the systems engineer.

Configuration Control and Management

Another place to go for an answer to the question posed in this section. The systems engineer can rely on a formality known as the configuration control board (CCB) for potential changes. After due explanation and consideration, the CCB will say either yea or nay to your request. Let's not make it more complicated than it needs to be, most of the time.

Iteration

Another related concept is that of "iteration" in systems engineering. In its broadest term, we have a notion of "iteration" in systems engineering that allows us to iterate until we find a satisfactory solution. This may also apply to very concrete situations, such as specific linear and non-linear equations [2].

Yet another context for the notion of iteration was discovered by this author in regard to a comment made by Mark Zuckerberg, head of Facebook. He was discussing problem-solving within the company and took note of one of their approaches. The key word was "iteration", as I recall. "You just keep iterating and eventually you wind up with an answer that works". Back in my school days, there was a set of words that we would use that meant pretty much the same. They were "you just kept grinding away". Apparently, there are many ways to say pretty much the same or similar things. But "iteration" still has a special place in the vocabulary of many, especially in the systems engineering world.

Yet another reference to "iteration" is found in the life cycle process models in the Systems Engineering Body of Knowledge. This is directly a part of the process models.

TBDs

Finally, we have the ubiquitous "TBD" (to be determined). This is used a lot when one reaches a point at which there is no good answer. We therefore put in a placeholder than temporizes. Let's wait a while for this answer, which will eventually reveal itself. Not a bad idea, based upon its history and use.

So the bottom line of the lesson learned in regard to this area goes something like this:

> in systems engineering projects and programs, there seem to be many ways to consider making constructive changes, to iterate so as to improve, but in the real world this is difficult to do once the configuration is well defined and documented. And don't forget the "TBDs" and the "iterations" when you're on the path, but not there yet.

References and Recommended Reading

1. Rechtin, E., "Systems Architecting", Prentice-Hall, 1991.
2. Kelley, C. T., "Iterative Methods for Linear and Nonlinear Equations", Society for Industrial and Applied Mathematics, 1995.

48. DOTSS

Developers, over the years, have created hundreds of systems of various types, typically sponsored (i.e., paid for) by the federal government. This is a huge market, and thousands of companies compete in it every year. The systems that have already been developed and installed are plentiful and of course are owned by the government. So one idea from this corner was to try to capture some of these developer-off-the-shelf systems (DOTSS) for multiple use. The conjecture was, and is, to gain some leverage and dramatically reduce the time and cost of creating "new" systems. It is the "reuse" of whole systems, if you will.

The cost and time considerations under DOTSS for a "new" system may be postulated, as an example, as some $100 million and five years. This is a "plain-vanilla" system that is amenable to a DOTSS approach. Using this approach, we conjecture that we will be able to "reuse" the system at one-fourth the cost ($40 million) and one-fifth the time (one year instead of five years). This results in a reduction to one-twentieth (a percentage of 200). Not bad for saving some cash and thinking outside the box. Worth an experiment or two, n'est-ce pas?

Over a period of about 24 months, this author and a representative of a leading high-tech company held a series of meetings in order to explain the nature of DOTSS. The final suggestion was to develop a study of DOTSS, to include candidates in DOTSS activities that could be replicated and reused at a quite low price. Some of the notions of the DOTSS briefing and paper were presented here [1,2].

The intent here is to proceed with some lessons learned with respect to the DOTSS notion. One of those lessons is that we may not have the incentive to save money that we think we have. Indeed, this idea was expressed at the Office of Management and Budget (OMB) level, where the government representation suggested that all program and project managers should spend all the monies that they are allocated. Saving money at that level is not a good approach for a variety of reasons.

Since DOTSS is a case of software reuse, it makes sense to take another look at the state of the art of this field. The previous look was some years ago [3] and needs to be updated.

References and Recommended Reading

1. Eisner, H., "Managing Complex Systems – Thinking Outside the Box", John Wiley, 2005.
2. Eisner, H., "The DOTSS Approach", PICMET Report, July 1997.
3. Gromadzki, R., "Extent and Issues of Software Reuse", PhD dissertation, The George Washington University, May 2004.

49. OBVERSITY

Here are 20 approaches to management that carry with them some possible lessons learned by considering the obverse:

1. Hire the smartest people, and then make sure not to listen to them as they provide their best analysis and advice for you
2. Meetings take managers away from their real work, so go easy on this commitment
3. People don't like bad news, so never report any to your boss
4. Spend lots of money on R & D, but don't pay any attention to their results and the implications
5. Never invest in your "cash cows"; you can keep milking them forever
6. No one in your organization knows as much as those you talk to that are outside your organization
7. Give your people unreasonable deadlines so as to constantly push their productivity
8. Control meetings by making sure that your ideas are the only ones presented and discussed
9. Decline to be a member of your boss's team since building your own team is your highest priority
10. Look as if you're listening to your teammate's argument but actually be preparing your next verbal assault

11. Keep your boss honest by challenging just about everything he or she says
12. Spend your whole budget on marketing; making the product is easy
13. After laboring over your strategic plan, ignore it for a year
14. Don't worry about industry trends; your company is the trendsetter
15. Add new-start initiatives every year – the more, the merrier
16. Keep your desktop clear of all but a few papers – a cluttered desk reveals a cluttered mind
17. Micromanage average or less than average performers to get them to produce according to company standards
18. Make decisions quickly so as to maintain forward momentum and respect
19. Never reengineer your business processes in house; farm it all out to the pros

50. VAILLANT, TURNED INTO LESSONS CONSIDERED

So as a final look at this issue I did an open-ended 50-year lessons learned query. It turned out that the result was less a matter of my specific systems background than it was an experience with a gent by the name of George Vaillant. Dr. Vaillant is well known for his "Harvard Study" during which he investigated success factors in a group of Harvard graduates [1]. So I did my own "Vaillant" study, but trying to glean success factors from a background in systems engineering. Here are the results, trusting how the mind works in its more mysterious ways.

So here's my last page look at this author's list of success factors (translate to lessons learned and attributes thereof), 100% based upon intuition and experience:

1. Grit
2. Intelligence
3. Listening
4. Focus
5. Integrity
6. Community involvement
7. Problem solver
8. Respect

9. Curiosity
10. Sense of humor
11. Resilient
12. Growth-oriented

Grit

Otherwise known as perseverance and a surprisingly hot topic these days [2]. And if you've seen the two movies (John Wayne, Jeff Bridges), you have to conclude that this attribute must be on your list of success factors. That's even if you don't use an eyepatch.

Intelligence

In this case, let's broaden the context to both the conventional IQ as well as emotional intelligence. This term relates to how well individuals deal with their emotions, including how well they recognize them. Indeed, it has been claimed that there are multiple intelligences, including linguistic, social, and inter-personal [3].

Listening

Who can deny that "listening" is a compelling attribute of the systems engineer? There is little that captures the mind and heart than this state of being – an engineer being truly interested in what a colleague has to say.

Focus

Goleman adds this to his amazing list of attributes that make it easier to understand how mind and intelligence work together [4]. And, according to Goleman, focus has three modes – orienting, selective attention, and open awareness.

Integrity

This is high on all lists – lack of integrity disqualifies!

Community Involvement

Connecting to one's community has now become a measure of success, and correctly so.

Problem Solver

Involvement is not enough. The next tangible step is to actually solve one or more problems and demonstrate the ability to do so.

Respect

Do you naturally garner the respect of others by your overall way of behaving and treating others?

Curiosity

Do you impress others by your interest in a wide variety of subjects – what are the key factors, and how do they work as well as interact? Do you ever find yourself in a library these days, even though Google is right there at your fingertips?

Sense of Humor

Deceivingly important in terms of positive reactions to those who display such an attribute. Does not mean crossing the boundary of not taking seriously enough.

Resilient

This is the ability to bounce back after a difficult negative experience. One might consider it the "slow-die" attribute in the context of system behavior.

Growth-Oriented

Orientation toward moving forward and in a positive direction.

References and Recommended Reading

1. Vaillant, G., "Triumphs of Experience – The Men of the Harvard Grant Study", Harvard University Press, 2012.
2. Duckworth, A., "Grit", Scribner, 2016.
3. Eisner, H., "'Thinking' – A Guide to Systems Engineering Problem Solving", CRC Press, 2019.
4. Goleman, D., "Focus – the Hidden Driver of Excellence", Harper, 2013.

Top Ten Lessons

6

1. STOVEPIPES

Sitting on top of the author's top ten list is the stovepipe issue. There appears to be a strong tendency for systems engineering management to try to integrate all stovepipes and thereby create an integrated system. In principle, we would like such a system, but we must be very careful in this domain. We need to recognize that many of the stovepipes have been developed with different types of software as well as languages. In this important aspect, they may not be integrable without an enormous effort with regard to both time and cost.

Based on this author's experience, the lesson learned is to back up a few steps and consider, in depth, the possible consequences of attempts at integration. This will require a deep look at the specific costs and effectiveness of such an integration activity. To reiterate – many stovepipe systems cannot easily be integrated due to this structure and software. In many situations, the cost-effective solution is to sit tight with a non-integrated system. This may be disappointing to many, but it may be the best approach. It also signals that if we want to integrate stovepipes, we need to take that into account before these stovepipes are produced. This early planning approach will likely lead to better results due to compatible structures and software. Remember, if a system is working, it's not a good idea, most of the time, to try to "fix" it (if it ain't broke).

2. MODELING AND SIMULATION

When and if it's possible to do so, building a model of a system is usually important and pays handsome dividends. It provides for the ability to "test" the system performance and to carry out tradeoff studies as a design activity. One type of "model" is the system prototype, which has come into greater acceptance in the last several years [1]. By way of illustration, three other specific "models" are briefly cited below.

The Parameter Dependency Diagram (PDD) [2]. This diagram shows dependencies between the key parameters of a system. It allows for a deeper understanding of how these parameters inter-relate, and sets the stage for a more penetrating quantitative analysis that will generally take more time and cost more.

The SystemiTool Systemigram [3,4]. This diagramming procedure is amply illustrated by Boardman and Sauser. As a diagram, it too shows relationships, but between features of systems and subsystems. The "gram" version is supported by the "tool" version so that there is ready access to the digital form of the tool.

Model-Based Systems Engineering (MBSE). Going back to basics, we have a fundamental approach to systems engineering in MBSE. Its underlying structure is a model, by definition. This approach has become widely accepted ever since its introduction many years ago [5].

3. ARCHITECTING

It's many years down the road, and we still do not have an agreed-upon procedure for architecting a system (this author's opinion). The need for such was recognized back in the 1990s by the DoD. That led to DoDAF, based initially upon the three-view concept, as follows:

a. An operational view
b. A systems view
c. A technical view

This author has presented an approach different from the above, which he called the Eisner Architecting Method (EAM) [6]. For those of a mind to do so, it can also be considered an Emergency Action Message. This approach shifts from a "views" notion to a "cost-effectiveness" idea. It also explicitly calls for a definition and evaluation of alternatives, leading to a preferred architecture among the alternatives.

Another aspect of this issue is software architecting. Despite the existence of several books on the matter [7,8], new work needs to be brought about with respect to the precise structure of software architecture and the compatibility between a system and software architecture. There are quite significant dividends that are likely to accrue from the correct research in this area.

One of the significant lessons learned here is that despite the various versions of DoDAF, it remains deficient in its basic structure and concept. Nonetheless we follow DoDAF mainly due to its history and DoD sponsorship. This lesson also has an elemental suggestion that another approach is called for.

4. AMID A WASH OF PAPER ...

Another important lesson is that systems engineering needs to become more like "a lean, mean fighting machine". Too much paper is produced that makes its practice, too often, burdensome and inefficient. Rechtin [9] points out, in his heuristics, that:

> amid a wash of paper a small number of documents become critical pivots around which every project's management revolves.

One can argue that this is at least partially related to these notions:

a. More agile processes
b. More rapid processes

The former is suggested, for example, by Turner and Boehm [10], and the latter by Eisner, Marciniak, and Pragluski [11]. In particular, the procedure for the just-named source has the name "RCASSE", or Rapid Computer-Aided System of Systems Engineering. Here we also have to point to the imperative: simplify, simplify, simplify. We are interested in less paper and more rapid production of the paper that's produced.

5. INDUSTRY INITIATIVES AND GOVERNMENT SUPPORT

We continue to learn the lesson that this field, as with many others, needs to be supported by research and people interactions. In this respect, one can single out two areas that reflect both, namely, that provided by INCOSE as well as the SERC (Systems Engineering Research Center) at the Stevens Institute of Technology. INCOSE is clearly paying attention to how to improve the overall field and is very successful at bringing together industry, government, and academia to develop priorities, share data, and solve problems that all appear to have in common. Large companies (as, for example, Lockheed Martin, General Dynamics, and Northrop Grumman) have internal programs that are critical to the field and also serve to keep the companies competitive in the SE marketplace. At the same time, the government has chosen to continue to support the SERC, a very good decision on their part. Conferences and publications (as per INCOSE's "Systems Engineering") are just a couple of concrete examples of how to support the overall field on a continuing basis.

6. THE ELDERS IN SYSTEMS ENGINEERING

Much wisdom is contained in the writings of the "elders" in the field of systems engineering. As a body of knowledge, we can look to the INCOSE Fellows to be a place to start. That brings one to the INCOSE website as well as "Google" searches. A first-order list of the recommended elders is provided below:

NAME	AREAS OF SPECIALIZATION
Eberhardt Rechtin	Systems Architecting and Management
Andrew Sage	Systems Engineering
Sarah Sheard	Complexity Analysis
Barry Boehm	Software Engineering and Metrics
Yacov Haimes	Risk Analysis and Mitigation

7. FUNCTIONAL DECOMPOSITION

This item finds its way into the top ten since it is an essential part of the suggested and new systems architecting process. The recommended architecture procedure [5] is to decompose to the third level:

a. The first level is the name of the system,
b. The second level constitutes the major system functions (often the subsystems), and

c. The third level decomposes to the sub-functions (often the sub-subsystems).

Further decomposition is not necessary, nor is it desirable. It generally leads to more non-productive activity. An exception is the system of systems. An example of such is the National Aviation System.

8. TEAM BUILDING

This treatise emphasizes the importance of the systems engineering team in terms of building a successful system. Some of the features of this approach are:

a. Maintain control over the selection of the team leader
b. Give the selected team leader sufficient responsibility and authority to get the job done
c. In particular, give the team leader sufficient time and budget to get the job done
d. Give special training to members of the team to understand how to execute as a high-performance team

We continue to have many examples of the need for team building as well as how they have been working over the years. To cite two, we see the Kelly team at Lockheed Martin and the Kelley team that has been leading-edge innovative with his company IDEO [12].

9. RISK ANALYSIS AND MITIGATION

We have learned that risk identification and mitigation is one of our most important activities as systems engineers. In this respect, we distinctly go beyond the analysis part to the mitigation part. Not enough time to make mitigation changes is not acceptable. Mitigating real system risk is the order of the day. This should require the re-design of portions of the system. This, in turn, is reflected in spending more of our budget and expanding our schedule. The urgency of this item is supported by the two catastrophic failures in Challenger and Columbia. Those realities force us to be realistic as well as decisive with our risk mitigation approach.

10. THE SYSTEMS APPROACH AND SYSTEMS THINKING

The very essence of systems engineering is to adopt and confirm a "systems" orientation and approach. We believe that this perspective leads us in the right direction and ultimately results in better systems of all shapes and sizes.

The origins of systems thinking and the systems approach can be seen in the work of Peter Senge [13] who called systems thinking the "Fifth Discipline". This seminal work has been widely accepted, with spinoffs that relate specifically to systems engineering.

References and Recommended Reading

1. Eisner, H., "Thinking – A Guide to Systems Engineering Problem Solving", CRC Press, 2019.
2. Eisner, H., "Computer-Aided Systems Engineering", Prentice-Hall, 1988.
3. "Applying Systems Thinking via Systemigrams TM for Defining the Body of Knowledge and Curriculum to Advance Systems Engineering (BKCASE) Project", Stevens Institute of Technology, Babbio Center, 5th Floor Castle Point on the Hudson, Hoboken, New Jersey.
4. Boardman, J., and B. Sauser, "Systems Thinking", CRC Press, 2008.
5. Wymore, A. W., "Model-Based Systems Engineering", CRC Press, 1993.
6. Eisner, H., "Systems Architecting – Methods and Examples", CRC Press, 2020.
7. Taylor, R., N. Medvidovic, and E. Dashofy, "Software Architecture", John Wiley, 2010.
8. Shaw, Mary, and D. Garlan, "Software Architecture – Perspectives on an Emerging Discipline", Pearson, 1996.
9. Rechtin, E., "Systems Architecting", Prentice-Hall, 1991.
10. Boehm, B., and R. Turner, "Balancing Agility and Discipline", Addison-Wesley, 2004.
11. Eisner, H., "Essentials of Project and Systems Engineering Management", Third Edition, John Wiley, p. 392.
12. Kelley, T., "The Ten Faces of Innovation", Currency Books, 2006.
13. Senge, Peter, "The Fifth Discipline – The Art and Practice of the Learning Organization", Doubleday, 1990.

Index

9 780367 422424